Revised Nuffield Advanced Science

General editor
Revised Nuffield
Advanced Chemistry
B. J. Stokes

Associate editor
inorganic chemistry
A. J. Furse

Associate editor
organic chemistry
M. D. W. Vokins

Associate editors
physical chemistry
**D. H. Mansfield,
Professor
E. H. Coulson,
Professor
Jon Ogborn**

Editor of this book
B. J. Stokes

Contributors to this book
A. W. B. Aylmer-Kelly
Professor E. H. Coulson
A. J. Furse
G. H. James
A. J. Malpas
D. H. Mansfield
Professor D. J. Millen
Professor Jon Ogborn
J. G. Raitt
B. J. Stokes
M. D. W. Vokins

Authors of Background reading
N. Coats
Dr A. W. L. Dudeney
Michael Fielder, Centrum Industrial Marketing Ltd
John S. Holman
Dr W. G. M. Jones, Imperial Chemical
Industries p.l.c., Pharmaceuticals Division
Dr C. Lowy
Dr J. F. Newman, Imperial Chemical
Industries p.l.c., Plant Protection Division
Jenny Salmon
Professor I. E. Smith
B. J. Stokes

Consultant on safety matters
Dr T. P. Borrows

CHEMISTRY STUDENTS' BOOK I

Topics 1 to 11

Revised Nuffield Advanced Science
Published for the Nuffield–Chelsea Curriculum Trust
by Longman Group Ltd

Longman Group UK Limited
Longman House, Burnt Mill, Harlow, Essex CM20 2JE, England
and Associated Companies throughout the world.

First published 1970
Revised edition first published 1984, reprinted with corrections 1984, 1985, 1987
Eighth impression 1990
Copyright © 1970, 1984 The Nuffield–Chelsea Curriculum Trust

Design and art direction by Ivan Dodd and Robin Hood
Illustrations by Rodney Paull and Gary Simmonds

Filmset in Times Roman and Univers
Produced by Longman Singapore Publishers Pte Ltd
Printed in Singapore

ISBN 0582 35361 0

Note
All references to the **Book of data** are to the **revised** edition
which is part of the present series.

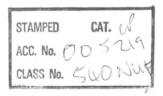
Cover picture
Magnesium sulphate (Epsom salt). Photomicrograph, taken in polarized
light, at ×8.
Copyright Paul Brierley.

Contents

Foreword

When the Nuffield Advanced Science series first appeared on the market in 1970, they were rapidly accepted as a notable contribution to the choices for the sixth form science curriculum. Devised by experienced teachers working in consultation with the universities and examination boards, and subjected to extensive trials in schools before publication, they introduced a new element of intellectual excitement into the work of A-level students. Though the period since publication has seen many debates on the sixth form curriculum, it is now clear that the Advanced Level framework of education will be with us for some years in its established form. Although various proposals for change in structure have not been accepted, the debate to which we contributed encouraged us to start looking at the scope and aims of our A-level courses and at the ways they were being used in schools. Much of value was learned during those investigations and has been extremely useful in the planning of the present revision.

The revision of the chemistry series under the general editorship of B. J. Stokes was conducted with the help of a committee under the chairmanship of Malcolm Frazer, then Professor of Chemical Education, University of East Anglia. We are grateful to him and to the committee. We also owe a considerable debt to the London Examinations Board which for many years has been responsible for the special Nuffield examinations in chemistry and to the subject officer, Peter Thompson, who has been an invaluable adviser on these matters.

The Nuffield–Chelsea Curriculum Trust is also grateful for the advice and recommendations received from its Advisory Committee, a body containing representatives from the teaching profession, the Association for Science Education, Her Majesty's Inspectorate, universities, and local authority advisers; the committee is under the chairmanship of Professor P. J. Black, academic adviser to the Trust.

Our appreciation also goes to the editors and authors of the first edition of Nuffield Advanced Chemistry, whose work, under the direction of E. H. Coulson, the project organizer, made this one of our most successful and influential ventures into curriculum development. Ernest Coulson's team of editors and writers included A. W. B. Aylmer-Kelly, Dr E. Glynn, H. R. Jones, A. J. Malpas, Dr A. L. Mansell, J. C. Mathews, Dr G. Van Praagh, J. G. Raitt, B. J. Stokes, R. Tremlett, and M. D. W. Vokins. A great part of their original work has been preserved in the new edition, on which several of them have acted as consultants.

I particularly wish to record our gratitude to Bryan Stokes, the General Editor of the revision. As a member of the original team he has an unrivalled understanding of the aims and scope of the first edition and as a practising teacher he possesses a particular awareness of the needs of pupils and teachers which has enriched the work of the revision. To him, to the editors working with him, A. J. Furse (Inorganic Chemistry), M. D. W. Vokins (Organic Chemistry), J. A. Hunt who is responsible for the Special Studies, and to the team responsible for the Physical Chemistry sections, Professor P. J. Black, J. Holman, D. H. Mansfield, Professor D. J. Millen, and Jon Ogborn, we offer our most sincere thanks.

I would also like to acknowledge the work of William Anderson, publications manager to the Trust, his colleagues, and our publishers, the Longman Group, for their assistance in the publication of these books. The editorial and publishing skills they contribute are essential to effective curriculum development.

K. W. Keohane,
Chairman, Nuffield–Chelsea Curriculum Trust

Introduction

This book

This book is volume I of the second edition of Nuffield Advanced Chemistry. Since the first edition was published, many developments in chemistry have taken place. New discoveries have been made; industrial processes have advanced; there have been changes in the way in which many chemicals are named, in the units in which many quantities are measured, in safety practices, and in a number of other matters. This second edition takes account of these.

Like its predecessor, this book contains the factual knowledge and the theoretical explanations that you will need; instructions for your practical work; background reading; and questions to test your understanding. But on its own it cannot give you a complete picture of chemistry at this level. To obtain this, you need to do the course, using all the other resources available to you – the *Book of data*, other textbooks, films and videorecordings, computer programmes, discussions with your teacher and with other students, and, above all, your own practical work.

This course

You are now beginning an advanced course in chemistry. What are you hoping to achieve by doing it? What is chemistry like at this level? Where will it get you? These are questions you can only answer at all fully after completing the course. But consider the essential nature of chemistry – what chemists do and how they think – and you will have some idea of what to expect.

Like all other sciences, chemistry is the study of the behaviour of materials. Behaviour implies change, and the type of change that interests you to a large extent influences the sort of science you do. Chemistry is the study of materials, but so is physics and so is engineering. What then makes chemistry distinctively chemical and different from physics or engineering?

Take a piece of aluminium as an example of a typical material. If you are interested, say, in the way in which the heat capacity or the electrical resistance of the aluminium changes when it is cooled to nearly the absolute zero of temperature, you are studying it as a physicist. If you are interested in the stresses which develop in a piece of aluminium when it forms part of a larger structure such as a supersonic aircraft, you may be studying the aluminium as an engineer. If you are interested in the changes that take place when the aluminium is heated in an atmosphere of chlorine, what happens to the aluminium atoms, how they react with the chlorine molecules, what energy changes accompany the reaction, and to what uses the product can be put, you are studying the aluminium as a chemist.

Physicists, engineers, metallurgists, and chemists, in studying materials such as aluminium, all share many common objectives and methods of working. Indeed, as time goes on the boundaries between chemistry, physics, engineering, metallurgy, etc. may disappear and the subjects may merge into one. At present, however, different scientists still have their own different emphases. Where does the emphasis lie in chemistry?

The entire physical world is composed of little more than one hundred different elements, but the atoms of these elements, in linking to form compounds, can combine in millions of different ways. Modern chemistry involves the study of the way the atoms are linked together by chemical bonds to form larger structures such as molecules. Much of chemistry is concerned with elucidating chemical structure and for this, many powerful methods such as molecular spectroscopy or X-ray crystallography are available to the chemist. But chemists are not only concerned with structure; they also study the changes which take place and the patterns of change when the atoms of a structure become disengaged from one another (the 'chemical bonds are broken') and link up to form new structures. This is the essence of a chemical, as opposed to a physical or any other type of process – the breaking and making of chemical bonds. Chemistry is the study of this bond breaking and bond making process in all its aspects, including how rapidly it occurs and what energy changes accompany it.

In understanding the underlying structure of a material we can often explain some of its macroscopic (large-scale) properties. For example, the hardness of diamond and the elasticity of rubber can be explained in terms of the ways in which the atoms and molecules, respectively, of these materials are arranged and interlinked. In understanding the bonding between atoms, and the material and energy changes that happen when bonds are made and broken, chemists can build structures to their own design to replace and often to supersede naturally occurring materials. This opens up exciting possibilities. Polymers for plastics, fertilizers for farming, alloys for aircraft construction, drugs to fight disease – these are a few examples of the contribution that chemistry can make to a better, longer life for us all.

Experiments and ideas: a simple example

For chemists, as for other workers in science, experiments and ideas go hand in hand. The two are very closely linked, as the following example shows.

Suppose you have been investigating the way in which hydrogen reacts with other chemical elements. After reading the reports of others who have worked in this field you have succeeded in making a small sample of lithium hydride by passing hydrogen gas over heated lithium metal, using an apparatus similar to that shown in figure i.

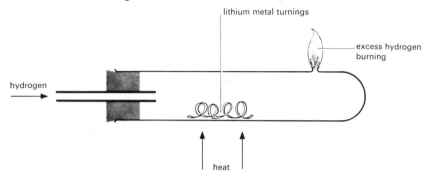

Figure i
Apparatus for making lithium hydride from lithium metal and hydrogen gas.

It called for considerable care and skill to cause a flammable gas to combine safely with a vigorous reactive metal. The compound is a white crystalline material, which reacts rapidly with water, so it must be kept in dry conditions. What would you do next? You have a sample of an interesting and (to you) new compound, lithium hydride. You could examine how it reacts in the presence of various laboratory reagents to see how your observations compare with those of others who have worked on lithium hydride. You could also follow up some of the many interesting questions which are raised as a result of the experiment you have just done. Here are a few.

A Could other hydrides be prepared in this way? Are there other ways of preparing hydrides? Are all hydrides similar in their behaviour or are there differences? Is there any pattern in the behaviour of hydrogen gas with other elements (e.g. the other alkali metals) or in the behaviour of the hydrides?

B What is happening at the level of the atoms and molecules during this reaction? What chemical bonds are broken and made and how much energy is required for this? What is the structure of lithium hydride? Is it ionic or covalent? How could this be settled?

C How rapidly did the reaction between hydrogen and lithium go? What would affect how fast it goes and in what way?

D Could we get the reaction to go backwards? What would be needed to do so? Was all the lithium converted to the hydride? Has this reaction 'gone to completion' or is it an equilibrium process?

E Does the reaction need heat to keep it going as opposed to starting it off? Is there any evidence of heat being produced by the reaction? If so, how much?

F Is lithium hydride likely to be of any use, for example in medicine or industry?

These groups of questions, which can apply to any chemical process you are studying, illustrate the main themes running through this course. They can be summarized as follows:

1 Particular chemical changes in materials.
2 Patterns in the chemical behaviour of materials.
3 Structure, including the structure of atoms and the structure of molecules and crystals.
4 Rates of reactions.
5 Equilibria in chemical systems.
6 Energy changes accompanying chemical changes.
7 Applications – industrial, medical, economic, and other social aspects.

Starting with a particular chemical change in a particular material, you are at once led to questions of patterns of chemical change, energy, structure, rates of reaction, equilibrium, and applications. These questions in turn will lead you back to experiment as a means of answering them. You may decide next, for example, to start a series of experiments to try to make other metal hydrides by this method. Or you may start an investigation of the factors influencing the rate of the chemical combination, and so on. Experiments lead to ideas and ideas to further experiments; together they lead to an increase in our understanding of the material universe.

The main point, then, about the activities of a chemist is that he deals in both ideas and experiments which go hand in hand; just as ideas without experiments are merely speculation, so experiments without ideas are merely trivial.

It is impossible to convey in a short section the great variety of interest and activity there is in modern chemistry. This you can find out only by *doing* chemistry, not by reading about it. Enough has been said, however, to make it clear that being 'good at' chemistry involves being able to report clearly and accurately what you do, what you observe, and what you think. This means developing your ability to write clear, unambiguous, readable English. Science is essentially an activity done by a community of people. This implies personal relationships. Good relationships depend on good communication.

Safety in practical chemistry

Finally, a note about safety. Many of the substances that are used in chemistry laboratories are potentially hazardous; corrosive, flammable, poisonous, or capable of reacting violently. Much of the equipment, too, has its dangers, whether from the sharp edges of a broken test-tube, or the risk of implosion when using a vacuum pump, to mention two examples. Provided that equipment and substances are handled with due attention to their inherent hazards – and that can only come from an understanding of the nature of these hazards – there is no reason why you should not complete your chemistry course in safety. Pay attention to any warnings given in the *Students' book*, on bottle labels, or by your teacher. Remember, too, that you have a responsibility for the safety of your fellow students, and the teachers and technicians – as they, in turn, have towards your safety. Bear this in mind when you clear away at the end of a practical session.

Some of the principal causes of accidents include:

carelessness (misreading bottle labels, for example);
use of excessive quantities of materials (10 g when the instructions
said 2 g);
lack of attention (failing to notice a tube becoming blocked until it is
too late, for example);
ignorance (if, for example, you had failed to check for hazards in advance).

Even when accidents do occur, their effects can be minimized by adopting suitable precautions:

wearing protective clothing (including eye protection);
using fume cupboards, and using safety screens, when appropriate;
knowing in advance what to do in an emergency;
never working alone in a laboratory.

Eye protection MUST be worn whenever there is a risk of damage to the eyes. In practice, this means that safety spectacles or goggles need to be worn for virtually all practical work in chemistry. It is absolutely essential for all operations in which substances are heated or gases generated, or in which acids, alkalis, or other corrosive materials are handled. Even if you yourself are not doing practical work, others in the laboratory may be, and protection should be worn for as long as any practical work is in progress. Those who normally wear spectacles will need goggles which can be worn over these.

As a chemist, by the end of this course you will be in an excellent position to understand – and warn others about – the hazards to be found not only in a laboratory, but generally, for example at home, or in a factory. This is one of the advantages – and responsibilities – of studying chemistry.

Contributors

Many people have contributed to this book. Final decisions on the content and method of treatment used in the first edition were made by the Headquarters team, who were also responsible for assembling and writing the material for the several draft versions that were used in school trials. The Headquarters team consisted of E. H. Coulson (organizer), A. W. B. Aylmer-Kelly, Dr E. Glynn, H. R. Jones, A. J. Malpas, Dr A. L. Mansell, J. C. Mathews, Dr G. Van Praagh, J. G. Raitt, B. J. Stokes, R. Tremlett, and M. D. W. Vokins.

The revision has been undertaken largely by three working groups, whose members were:

Inorganic chemistry: A. J. Furse (chairman), K. W. Badman, M. C. V. Cane, C. Nicholls, and D. Russell.

Organic chemistry: M. D. W. Vokins (chairman), J. J. Eggleton, G. H. James, and Professor D. J. Waddington.

Physical chemistry: Professor M. J. Frazer (chairman), Professor P. J. Black, Dr T. P. Borrows, John S. Holman, D. H. Mansfield, Professor D. J. Millen, and Jon Ogborn.

Advice on safety matters has been given by Dr T. P. Borrows, Chairman of the Safety Committee of the Association for Science Education.

The authors of Background reading are listed earlier in these preliminary pages, but we should also like to thank those who assisted as consultants and helped to bring the material up to date. They are F. L. Hodges, Public Affairs and Information Department, British Petroleum Company p.l.c.; A. J. Rooke, BP Educational Service; Jenny Salmon; Dr T. J. Veasey; Dr T. J. Walker, Imperial Chemical Industries p.l.c., Petrochemicals and Plastics Division; and J. A. E. White, Imperial Chemical Industries p.l.c., Pharmaceuticals Division. All the infra-red spectra were kindly supplied by Reuben B. Girling, University of York.

This book has benefited greatly from the valuable help and advice that have been generously given by teachers in schools and in universities and other institutions of higher education. In particular, the comments and suggestions of teachers taking part in the school trials, both of the original course, and of the revised Topics, have made a vital contribution to the final form of the published material.

Finally, as editor, I should like to record my thanks to the Publications Department of the Nuffield–Chelsea Curriculum Trust for their help, and particularly to Mary de Zouche for her meticulous and painstaking attention to detail in the preparation of the manuscript for publication and to her colleagues, Deborah Williams, Hendrina Ellis, and Nina Konrad.

B. J. Stokes

TOPIC 1

The Periodic Table 1: an introduction

1.1
AMOUNT OF SUBSTANCE

One fundamental quantity chemists need to know when measuring out materials is the amount of chemical substance.

In physics it is sometimes sufficient to measure quantities of material in units of mass (grams or kilograms), for example in considering the motions of bodies. For chemical purposes this is not sufficient because proper chemical comparisons between different materials can only be made if comparable numbers of particles are considered in each case.

Suppose we have one gram of hydrogen available for combination and a supply of the metals lithium, Li, sodium, Na, and potassium, K. Assuming that all the hydrogen can be efficiently converted into hydride and that the formulae of all three hydrides are of the form MH, using which metal could we obtain the largest amount of hydride?

From one gram of hydrogen the masses of the three hydrides obtainable are 8 g LiH, 24 g NaH, and 40 g KH, so that at first sight it may seem that with potassium the largest amount of hydride is obtained and with lithium the least. It is clear, however, that in the three equations,

$$Li(s) + \tfrac{1}{2}H_2(g) \longrightarrow LiH(s)$$
$$Na(s) + \tfrac{1}{2}H_2(g) \longrightarrow NaH(s)$$
$$K(s) + \tfrac{1}{2}H_2(g) \longrightarrow KH(s)$$

for every atom of hydrogen reacting, one atom of metal reacts and one formula unit of the hydride is formed. Therefore there is a sense in which the same amount of metal reacts with one gram of hydrogen in each case to give the same amount of hydride.

For chemists, equal amounts of different substances mean equal numbers of particles. So to compare equal amounts we have to weigh out, not equal masses, but masses of substances which are in the ratio of the masses of the particles present in them. The fact that the atoms of different elements have different masses is in a sense an inconvenience.

It was during the early part of the nineteenth century that chemists began to realize that what mattered to them was *numbers of particles* of substances. Hydrogen, as the lightest element, was chosen as a reference element and the mass of the hydrogen atom was chosen as the unit of atomic mass. Masses of other atoms and molecules were expressed in terms of the number of times they were as heavy as one hydrogen atom.

Since the unit of mass was the gram it was natural to choose one gram of hydrogen as a unit amount of substance. This unit amount of substance later came to be called one mole of hydrogen atoms. The number of hydrogen atoms in one gram of hydrogen was called first the Avogadro number, and later the Avogadro constant was introduced.

New definitions of relative atomic mass, the mole, and the Avogadro constant have since been introduced, and the modern versions are given later in this section. You may find it useful to write each definition in your notebook as you come to it.

Definition of relative atomic mass

In 1961 a new standard of atomic mass was adopted, based on the isotope of carbon of mass twelve (referred to as carbon-12, or ^{12}C). The change was made for a number of reasons, mostly concerned with the existence of isotopes, and as carbon atoms have a mass about 12 times as great as that of hydrogen the change of definition caused very little change in the numerical values.

The *relative atomic mass* of an element is now defined as the mass of one atom of the element on a scale chosen so that the mass of one atom of the ^{12}C isotope of carbon is 12 units exactly.

A table of relative atomic masses is to be found in the *Book of data*.

Definition of the mole

The mole (unit amount of substance) is also defined in terms of the ^{12}C isotope of carbon.

One *mole* of any substance is the amount of substance which contains as many elementary units as there are atoms in 12 grams (exactly) of pure carbon-12. The elementary unit must be specified and may be an atom, a molecule, an ion, an electron, etc., or a group of such entities.

The mole is one of the seven basic units of measurement of the International System of Units (SI units). A list of all of these, together with their definitions, is given in the *Book of data*.

Notice that the definition of the mole refers to elementary units (molecules, atoms, ions, etc.), and that it is vital when referring to amounts of substances to specify what elementary units are meant. This is because a phrase such as 'one mole of chlorine' is ambiguous: it could mean one mole of chlorine molecules (Cl_2), that is, 71.0 g, or it could mean one mole of chlorine atoms (Cl), that is 35.5 g. For this reason the formula of the substance being considered must always be stated, and the correct phrases should be 'one mole of chlorine molecules Cl_2,' and 'one mole of chlorine atoms Cl'.

In substances consisting of giant lattices, whether they are covalent such as silicon dioxide, SiO_2, or ionic, such as sodium chloride, Na^+Cl^-, the formula specifies the elementary units to which the mole refers. Notice that, for example, one mole of sodium chloride, NaCl, contains one mole of sodium ions and one mole of chloride ions, whereas one mole of calcium chloride, $CaCl_2$, consists of one mole of calcium ions, Ca^{2+}, and two moles of chloride ions Cl^-. Difficulties which might arise are dealt with by stating the formula being considered in every case.

If we want to calculate the amount of a substance in moles, knowing its mass in grams, we must divide the mass by the mass in grams of one mole of particles of the substance concerned. The relation is

$$\frac{\text{amount}}{\text{in moles}} = \frac{\text{mass in grams}}{\text{mass of one mole in grams per mole}}$$

Dissolving a mole of a substance in sufficient water to make 1 cubic decimetre (litre)* of solution gives a solution of concentration one mole per cubic decimetre, sometimes known as a *molar solution*. A solution of sodium chloride of concentration $1\,mol\,dm^{-3}$, for example, is made by dissolving one mole of sodium chloride, NaCl (58.5 g), in water and making the solution up to a total volume of $1\,dm^3$. The abbreviation 'M' is sometimes used to indicate this concentration; a solution of sodium chloride of concentration $1\,mol\,dm^{-3}$ may be referred to as M sodium chloride solution. A solution of concentration $2\,mol\,dm^{-3}$ is referred to as '2M' and so on.

From the above considerations you can see that a solution may have a concentration of $1\,mol\,dm^{-3}$ with respect to the substance dissolved, but $2\,mol\,dm^{-3}$ with respect to one of its ions. For example, a solution of calcium chloride of concentration $1\,mol\,dm^{-3}$ has a concentration of $1\,mol\,dm^{-3}$ with respect to calcium ions, but of $2\,mol\,dm^{-3}$ with respect to chloride ions. Because of this, solutions may sometimes be referred to as, for example, '$2\,mol\,dm^{-3}$ with respect to Al^{3+}', if the nature of the other ion is irrelevant to the use to which the solution is to be put.

Definition of the Avogadro constant

The mole, which is the unit of *amount of substance*, is defined in terms of a number of elementary particles. The Avogadro constant is the constant of

---

* The cubic decimetre, symbol dm^3, is now the preferred international name for the unit of volume otherwise known as the litre. In this book, when referring to volumes, we shall normally use cubic centimetres, cm^3 (and not millilitres, ml) and cubic decimetres, dm^3 (and not litres, l). You may find the alternative names in other books.

proportionality between amount of substance and number of specified particles of that substance. It is represented by the symbol L, and has a numerical value of 6.02×10^{23}; its unit is mole^{-1}.

$$\begin{array}{ccc} \text{amount} \\ \text{in mol} \end{array} \times \begin{array}{c} \text{Avogadro constant} \\ \text{mol}^{-1} \end{array} = \begin{array}{c} \text{number of} \\ \text{specified particles} \end{array}$$

The Avogadro constant can be determined experimentally in a number of ways. The most accurate so far known depends upon the X-ray measurement of the internuclear distances of atoms or ions in crystals, and is described in Topic 7. A radioactivity method is described here, and an electrical method forms the basis of the experiment which follows.

Radioactivity method. When the radioactive element radium decays, each atom emits an α-particle. These particles can be collected in a suitable container, where they pick up stray electrons and become helium atoms. The volume of the trapped helium can be measured. It is very small, but measurable if the collection is continued over a long period. A typical figure is 0.043 cm^3 per gram of radium per year.

The number of α-particles emitted can be counted by means of a Geiger counter. The fraction emitted in a small solid angle is measured, and then multiplied to find the total number emitted; this is very large, a typical value being 11.6×10^{17} per gram of radium per year.

These figures tell us that 0.043 cm^3 of helium consists of 11.6×10^{17} atoms. One mole of helium atoms, which occupies 22.4 cubic decimetres at a standard temperature and pressure (s.t.p.), therefore contains

$$\frac{11.6 \times 10^{17} \times 22\,400}{0.043} = 6.04 \times 10^{23} \text{ atoms}$$

EXPERIMENT 1.1
Electrical method of finding a value for the Avogadro constant

In this experiment you are going to electrolyse copper(II) sulphate solution. You will measure the current that flows, and the time for which it flows, and thus work out the number of coulombs that have passed; and you will find the increase in mass of the cathode to determine the mass of copper that has been deposited, and the decrease in mass of the anode to determine the mass of copper that has been taken into solution.

Assuming a knowledge of the relative atomic mass of copper (63.5), the charge on the copper ion $(+2)$, and the charge on the electron $(1.60 \times 10^{-19}$ C), you can then work out a value for the Avogadro constant, L.

Note on recording the results of your practical work. Your teacher will probably give you some advice on how to write an account of your practical work in your notebook. It is important that you keep some record of the way each experiment is done, how any calculations are made, what results are obtained, and what conclusions are reached. This information will be needed to complement the contents of this book.

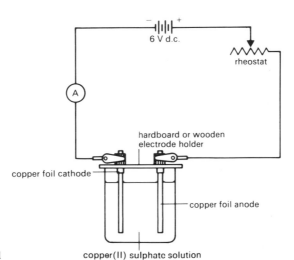

Figure 1.1   copper(II) sulphate solution

Procedure

The completed apparatus for this experiment is shown in figure 1.1. Assemble it in the following order:

1 Carefully clean the copper foil cathode and anode by rubbing them with steel wool, wash them with water and then wash them twice with propanone, wave them in the air for a minute or two to evaporate most of the propanone, and dry by warming high above a Bunsen flame. *TAKE CARE:* remember that propanone is highly flammable. Mark the upper end of the copper foil cathode, in pencil, with a C; mark the upper end of the copper foil anode with an A.

2 Find the masses of copper cathode and anode, and enter them in your notebook. A suitable table for this is given below.

3 Three-quarters fill the beaker with copper sulphate solution and insert the copper electrodes, using the holder provided.

4 Complete the rest of the circuit, using connecting wires and crocodile clips, but do not attach the crocodile clip on the wire from the ammeter to the copper cathode. Check the rest of the connections, making sure that the copper cathode and anode are in their correct positions.

5 Attach the crocodile clip to the copper cathode, start the stop clock (or note the time on a clock with a seconds hand), and adjust the rheostat to give a current of 0.20 ampere.

Allow the current to pass through the solution for 20 minutes, keeping it at 0.20 A by adjusting the rheostat when necessary. At the end of 20 minutes disconnect the two copper electrodes and rinse them with water. The anode (A) will need a strong jet of water from the tap, followed by a firm wipe with a paper tissue to remove the film which collects on the surface. Rinse each electrode twice with propanone, and dry as before. Find the mass of each electrode separately and record the results in your notebook, in a table similar to the one that follows.

	Copper cathode /g	Copper anode /g
Initial mass		
Final mass		
Change in mass		

Current used ..... amperes

Time current passed ..... seconds

Calculation

1 Work out the quantity of electricity (coulombs) passed through the solution.

Quantity (coulombs) = current (amperes) × time (seconds)

2 Calculate the quantity of electricity required to deposit 63.5 g copper.

3 The charge on an electron is 1.60×10^{-19} coulomb. Calculate the number of electrons required to discharge 63.5 g of copper ions.

4 Each copper ion requires 2 electrons for discharge

$$Cu^{2+}(aq) + 2e^- \longrightarrow Cu(s)$$

How many copper ions are in 1 mole (63.5 g)? This is the Avogadro constant.

Compare the results that you obtain, using both the increase in mass of the cathode and the decrease in mass of the anode.

1.2
THE PERIODIC TABLE

If elements are arranged in order of increasing relative atomic mass, elements having similar properties recur at periodic intervals in this list.

This *periodicity* is particularly well seen if the elements are arranged in the form of a table. The first successful table to group elements according to their chemical behaviour in this way was devised by Mendeleev, the great Russian chemist, in 1869. Mendeleev based his table on the sixty-odd elements then known. Since that time the table has grown to accommodate over one hundred elements, and has been rearranged to take account of the electronic structures of the atoms, which were quite unknown to Mendeleev. That the original concept has proved capable of absorbing this new knowledge, however, shows the correctness of Mendeleev's original proposal.

Now known as the *Periodic Table*, this classification of elements is one of the great achievements of chemical science. The history of its development is outlined in the Background reading at the end of this section.

Have a look at a modern Periodic Table; one is given inside the back cover of the *Book of data*. A horizontal row across the Table is known as a *period*. Periods are numbered from the top downwards. Period 1 thus consists of hydrogen and helium; the elements lithium (Li) to neon (Ne) form period 2, and so on.

A vertical column is called a *group*. Groups having elements in periods 2 and 3 are numbered from left to right, with the exception of the group headed by neon, which is called Group 0. Besides numbers, several groups have names.

Group number	Name
I	alkali metals
II	alkaline earth metals
VII	halogens
0	noble gases

Three horizontal regions of the Table have names, and these are the transition elements, the lanthanides, and the actinides. The *transition elements* are those in the groups headed by titanium (Ti) to copper (Cu) inclusive. The *lanthanides* are the elements cerium (Ce) to lutetium (Lu) inclusive. Lanthanum and the lanthanides are sometimes known as the rare earth elements. The *actinides* are the elements thorium (Th) to lawrencium (Lw) inclusive.

These names are indicated on the outline Periodic Table (figure 1.2).

If the elements are numbered along each period from left to right, starting at period 1, then period 2, and so on, the number given to each element is called its *atomic number*. This number has a greater significance for the Periodic Table than has relative atomic mass, as will be seen later.

Element	Li	Be	B	C	N	O	F	Ne
Structural type	← giant lattices →				← molecules →			
Melting point /°C	181	1278	2300	3550	−210	−218	−220	−249
Heat of fusion/kJ mol⁻¹	3.0	12.5	22.2	—	0.36	0.22	2.55	0.34

Element	Na	Mg	Al	Si	P (white)	S	Cl	Ar
Structural type	← giant lattices →				← molecules →			
Melting point/°C	98	649	660	1410	44	119	−101	−189
Heat of fusion/kJ mol⁻¹	2.6	8.9	10.7	46.4	0.63	1.41	3.2	1.2

Table 1.1
(In this table and in the *Book of data*, the mole consists of single atoms. Care is needed in interpreting the results for gases with diatomic molecules.)

the presence of small molecules that are easily separated from one another.

Draw up a similar table in your notebook showing the boiling points and heats of vaporization of these elements; values can be obtained from the *Book of data*. Do they follow a similar pattern to that seen in table 1.1? If so, make a note of what that pattern is.

BACKGROUND READING
The early history of the Periodic Table

The Periodic Table is one of the great achievements of chemical science, as it brings order and system to the enormous amount of information which is available about the chemical elements and their compounds. Before the Periodic Table was suggested, several scientists had made attempts to classify the elements according to their properties. Three such attempts were those of

Antoine Lavoisier with his 'Groups',
J. W. Döbereiner with his 'Triads', and
John Newlands with his 'Octaves'.

Order in Groups: Lavoisier

Lavoisier, a French nobleman, had many scientific interests. In 1790 he was a member of the commission that introduced the metric system, but he is most famous for his explanation of burning which led to the downfall of the 'phlogiston theory'. In 1789 Lavoisier published one of the most influential books on chemistry ever written. It was called *Traité Elémentaire de Chimie* (*Elements of Chemistry*), and in it he gave a list of 'simple substances not decomposed by any known process of analysis', or, as we would say, a list of 'the elements'. He divided this list into several groups, based on the similar chemical behaviour of the elements in each group. As you can see from figure 1.5 (photographed from his book), he put oxygen, nitrogen, hydrogen, light, and heat together in

Figure 1.4
Antoine Lavoisier (1743–94), who first attempted to sort the elements into groups.
Photograph, Ann Ronan Picture Library.

the first group. In the second, he put sulphur, phosphorus, carbon, chlorine, and fluorine. He called these the 'acidifiable' elements, by which he meant those elements that formed an acid on combining with oxygen. In the third group, he put the metals: silver, arsenic, bismuth, cobalt, copper, tin, lead, tungsten, and zinc. Finally, in the fourth group, he put what he called the 'simple earthy salt-forming substances': lime (calcium oxide), baryta (barium oxide), magnesia (magnesium oxide), alumina (aluminium oxide), and silica (silicon dioxide). In Lavoisier's time, this last group was believed to be composed of elements because the substances had not then been broken down into anything simpler. We now know them to be compounds which are very difficult to decompose into their constituent elements.

Order in Threes: Döbereiner's Triads

Lavoisier's work was an important beginning; it implanted the idea of a relationship between the elements, but it didn't give much of a clue to the eventual

	Noms nouveaux.	*Noms anciens correspondans.*
	Lumière.........	Lumière₁
Substances sim- *ples qui appar-* *tiennent aux* *trois règnes &* *qu'on peut regar-* *der comme les* *élémens des* *corps.*	Calorique........	Chaleur. Principe de la chaleur. Fluide igné. Feu. Matière du feu & de la chaleur.
	Oxygène.........	Air déphlogiſtiqué. Air empiréal. Air vital. Baſe de l'air vital.
	Azote...........	Gaz phlogiſtiqué. Mofete. Baſe de la mofete.
	Hydrogène.	Gaz inflammable. Baſe du gaz inflammable₁
Substances sim- *ples non métalli-* *ques oxidables &* *acidifiables.*	Soufre...........	Soufre.
	Phoſphore........	Phoſphore.
	Carbone.........	Charbon pur₁
	Radical muriatique.	Inconnu.
	Radical fluorique..	Inconnu.
	Radical borracique..	Inconnu.
Substances sim- *ples métalliques* *oxidables & aci-* *difiables.*	Antimoine........	Antimoine₁
	Argent...........	Argent.
	Arſenic; . ;	Arſenic.
	Biſmuth..........	Biſmuth₁
	Cobolt...........	Cobolt.
	Cuivre..........	Cuivre.
	Etain............	Etain.
	Fer.	Fer.
	Manganèſe.	Manganèſe.
	Mercure..........	Mercure.
	Molybdène.......	Molybdène₁
	Nickel...........	Nickel.
	Or..............	Or.
	Platine..........	Platine.
	Plomb...........	Plomb.
	Tungſtène........	Tungſtène.
	Zinc..	Zinc.
Substances sim- *ples salifiables* *terreuses.*	Chaux...........	Terre calcaire, chaux.
	Magnéſie.........	Magnéſie, baſe du ſel d'Epſom.
	Baryte...........	Barote, terre peſante₁
	Alumine.........	Argile, terre de l'alun, baſe de l'alun.
	Silice...........	Terre ſiliceuſe, terre vitrifiable.

Figure 1.5
A page from Lavoisier's famous book *Traité Elémentaire de Chimie*, showing the way in which he grouped the chemical substances then known. Several of the substances he listed were not, in fact, elements. Can you identify these?
Photographed by courtesy of the Royal Society.

Figure 1.6
J. W. Döbereiner (1780–1849) who grouped some of the elements in 'triads' (groups of three).

pattern. The next step in piecing together the puzzle was taken in 1817 by a German scientist, J. W. Döbereiner, who was a professor at the University of Jena where Goethe, the German poet, attended his lectures.

Döbereiner realized that three recently isolated elements, calcium, strontium, and barium, all had properties that were strikingly similar. Calcium, strontium, and barium all occur naturally as carbonates and sulphates which do not dissolve in water and which do not decompose easily when heated. Their chlorides are all soluble in water and their oxides dissolve in water to produce a strongly alkaline solution. The three elements were also isolated in the same way: electrolysis of the molten chlorides by Davy in 1808.

Döbereiner noticed that the relative atomic mass of strontium (88) was almost midway between the relative atomic masses of calcium (40) and barium (137). He called this group of three elements a 'triad'. In later years, he noticed also that two other 'triads' of elements – chlorine, bromine, and iodine, and lithium, sodium, and potassium – repeated the same pattern. Not only were their properties similar, but also the relative atomic mass of the middle one fell halfway between those of the other two. Döbereiner thought he had discovered the key to the jigsaw: the elements of nature fitted together in threes. His discovery became known as the 'Law of Triads'. But this grouping in threes was restricted to only a few elements. What of all the others? Döbereiner's observation that the link-up between the elements depended in some way upon their relative atomic masses provided the key.

Order in Eights: Newlands's Octaves

Before further progress could be made, it was necessary to find the relative atomic masses of all the known elements with some degree of accuracy. In Döbereiner's time relative atomic masses were still largely a matter of guesswork, and it was not until 1857, after intensive work by other scientists, that an accurate method was found for determining them. Seven years later John Newlands, a British chemist, found that when the elements were arranged in order of their relative atomic mass, with hydrogen (the lightest element) numbered 1, the second lightest (thought at the time to be lithium) numbered 2, the third numbered 3, and so on, then elements 1, 8, and 15 were similar, as were elements 2, 9, and 16 and so on. As Newlands wrote, 'the eighth element, starting from a given one, is a kind of repetition of the first, like the eighth note in an octave of music'. This kind of repetition, with similar properties 'periodically' recurring, is called periodic, and is the origin of the name 'periodic table'. Unfortunately, although the periodic relationship Newlands had found held good for the first sixteen elements, it did not work after the seventeenth. This made scientists rather reluctant to accept Newlands's ideas.

The following is an extract from the report of a meeting of the Chemical Society on 1 March 1866, at which Newlands announced his observations on the periodicity of properties of the elements. It will be seen from the report that his ideas met with a good deal of scepticism.

'Mr John A. R. Newlands read a paper entitled "The law of Octaves, and the Causes of Numerical Relations among the Atomic Weights". The author claims the discovery of a law according to which the elements analogous in their properties exhibit peculiar relationships, similar to those subsisting in music between a note and its octave. Starting from the atomic weights on Cannizzaro's system, the author arranges the known elements in order of succession, beginning with the lowest atomic weight (hydrogen) and ending with thorium (= 231.5): placing, however, nickel and cobalt, platinum and iridium, cerium and lanthanum, etc., in positions of absolute equality or in the same line. The fifty-six elements so arranged are said to form the compass of eight octaves, and the author finds that chlorine, bromine, iodine, and fluorine are thus brought into the same line, or occupy corresponding places in his scale. Nitrogen and phosphorus, oxygen and sulphur, etc., are also considered as forming true octaves.

'The author's supposition will be exemplified in [a table] shown to the meeting, and here subjoined.

H	1	F	8	Cl	15	Co and Ni	22	Br	29	Pd	36	I	42	Pt and Ir	50
Li	2	Na	9	K	16	Cu	23	Rb	30	Ag	37	Ca	44	Os	51
G	3	Mg	10	Ca	17	Zn	24	Sr	31	Cd	38	Ba and V	45	Hg	52
Bo	4	Al	11	Cr	19	Y	25	Ce and La	33	U	40	Ta	46	Tl	53
C	5	Si	12	Ti	18	In	26	Zr	32	Sn	39	W	47	Pb	54
N	6	P	13	Mn	20	As	27	Di and Mo	34	Sb	41	Nb	48	Bi	55
O	7	S	14	Fe	21	Se	28	Ro and Ru	35	Te	43	Au	49	Th	56

Table
Elements arranged in octaves.

'Dr Gladstone made objection on the score of its having been assumed that no elements remain to be discovered. The last few years had brought forth thallium, indium, caesium, and rubidium, and now the finding of one more would throw out the whole system. The speaker believed there was as close an analogy subsisting between the metals named in the last vertical column as in any of the elements standing on the same horizontal line.

'Professor G. F. Foster humorously inquired of Mr Newlands whether he had ever examined the elements according to the order of their initial letters? For he believed that any arrangement would present occasional coincidences, but he condemned one which placed so far apart manganese and chromium, or iron from nickel and cobalt.

'Mr Newlands said that he had tried several other schemes before arriving at that now proposed. One founded upon the specific gravity of the elements had altogether failed, and no relation could be worked out of the atomic weights under any other system than that of Cannizzaro.'

Meyer's Curves

Meanwhile, two other chemists had been grappling with the same problem that Newlands had attempted to solve. One, Lothar Meyer, was working in Germany; and the other, Dimitri Mendeleev, in Russia. In 1864 Lothar Meyer, then thirty-four years old and Professor of Chemistry at Tübingen, worked out the volume that one mole of atoms of an element would occupy if it were a solid. This he called the 'atomic volume' of the element. He plotted atomic volumes against relative atomic masses to give a result similar to the bar chart in figure 1.3.

From Lothar Meyer's curve it is possible to arrive at a periodic arrangement of the elements similar to that put forward by Mendeleev (as discussed below); indeed Meyer did produce such a table. But most of the credit for this arrangement of the elements goes to Mendeleev because he was a man bold enough to make some detailed predictions about elements which nobody had yet discovered.

Figure 1.7
D. Mendeleev (1834–1907), who was the first to produce a periodic table similar
to those in use today.
Photograph, BBC Hulton Picture Library.

Mendeleev's Table

Dimitri Mendeleev, who published his work in 1869, was Professor of Chemistry
at St Petersburg (now Leningrad). He arranged the elements according to their
relative atomic masses, much as Newlands had done, but with two important
differences: he left gaps for elements which, he said, had not yet been discovered;
and he listed separately some 'odd' elements (for example, cobalt and nickel)
whose properties did not fit in with those of the main groups. This regrouping
helped to remove the obstacle to the use of Newlands's arrangement and, apart
from the fact that it contained only about sixty elements, Mendeleev's periodic
table is in principle much the same as that which we use today. In other words,
the outline of the jigsaw was complete, although a number of the pieces were
still missing.

Perhaps the most important feature of Mendeleev's work was that he left
gaps in his table where he thought the 'missing' elements should be. This was

important because, if a theoretical idea in science is to be really useful, it should not only explain the known facts but also enable new things to be predicted from it. In this way the theory can be tested by seeing whether or not the predictions prove to be correct, and also the theory can lead to scientific advance from following up the new ideas. Both Newlands and Lothar Meyer failed to provide a basis for prediction in their work. But with Mendeleev – and this drew attention to his table in the first place – not only were elements discovered which fitted the gaps in the table that he had left for them, but also their properties agreed remarkably well with those that Mendeleev had said they should have.

Take one example. When Mendeleev was arranging his table, he left a gap for an element between silicon and tin. He predicted that the relative atomic mass of this element would be 72 and its density 5.5 – basing his predictions on the properties of other known elements which surrounded the gap. Fifteen years later the element was discovered. It had a relative atomic mass of 72.6 and a density of 5.35. It was given the name germanium. Mendeleev made other predictions about it too. Table 1.2 shows how closely he was able to predict the properties of this new element, and provides confirmation of the correctness of his ideas.

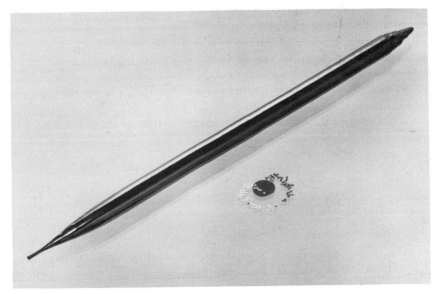

Figure 1.8
A long crystal of the element, germanium, the properties of which Mendeleev predicted so accurately. The crystal is shown approximately a quarter the actual size, with a few flakes of the element in the foreground.
Photograph, ITT.

Mendeleev's predictions	Observed properties
Colour will be light grey	Colour is dark grey
Will combine with two atoms of oxygen to form a white powder (the oxide) with a high melting point	Combines with two atoms of oxygen to form a white powder (the oxide) with a melting point above 1000 °C
The oxide will have a specific gravity of 4.7	Specific gravity of the oxide is 4.228
The chloride will have a boiling point of less than 100 °C	The chloride boils at 84 °C
The specific gravity of the chloride will be 1.9	The specific gravity of the chloride is 1.844

Table 1.2
Some properties of germanium.

SUMMARY

At the end of this Topic you should:

1 know the definition of relative atomic mass, the mole, and the Avogadro constant;

2 be able to do calculations using moles, and solutions of concentrations expressed in $mol\,dm^{-3}$ (molar solutions);

3 know how to find an experimental value for the Avogadro constant;

4 be familiar with the Periodic Table, including the names given to the various named groups and periods;

5 know that the atomic number of an element gives the order in which the element appears in the Periodic Table;

6 understand the meaning of periodicity, and be aware of the periodicity of some physical properties of the elements when arranged in atomic number sequence;

7 be aware of some aspects of the history of the Periodic Table.

PROBLEMS

* Indicates that the *Book of data* is needed.

*1 Instructions for practical work often give the quantities of reactants in terms of moles. But for the actual measurement of these quantities you need to convert them into the units of the measuring instrument; for example, grams, cm^3, etc. The following questions require you to convert molar quantities in this manner with the aid of your *Book of data*.

a What is the mass of 0.1 mole of zirconium atoms, Zr?

b What is the mass of 0.02 mole of thorium atoms, Th?

c What is the mass of 2 moles of nickel atoms, Ni?

d What is the mass of 1 mole of phosphorus molecules, P_4?
e What is the mass of 0.5 mole of sodium chloride, NaCl?
f What is the volume of 0.1 mole of mercury atoms, Hg?
g What is the volume of 1 mole of sulphuric acid, H_2SO_4?
h What is the volume of 0.2 mole of the alcohol ethanol, C_2H_5OH?
i What mass of anhydrous calcium chloride, $CaCl_2$, contains 2 moles of chloride ions, Cl^-?
j What mass of aluminium sulphate, $Al_2(SO_4)_3$, contains 1 mole of aluminium ions, Al^{3+}?

*2 Calculate the mass of each of the following:

a 1 mole of hydrogen molecules, H_2
b 1 mole of hydrogen atoms, H
c 1 mole of silica, SiO_2
d 0.5 mole of carbon dioxide, CO_2
e 0.25 mole of hydrated sodium carbonate, $Na_2CO_3 \cdot 10H_2O$

*3 How many moles or part of a mole are each of the following?

a 32 g of oxygen molecules, O_2
b 32 g of oxygen atoms, O
c 31 g of phosphorus molecules, P_4
d 32 g of sulphur molecules, S_8
e 50 g of calcium carbonate, $CaCO_3$

*4 What mass of each of the following is dissolved in 250 cm^3 of a solution of concentration 0.100 mol dm^{-3}?

a Hydrochloric acid, HCl
b Sulphuric acid, H_2SO_4
c Sodium hydroxide, NaOH
d Potassium manganate(VII), $KMnO_4$
e Sodium thiosulphate, $Na_2S_2O_3 \cdot 5H_2O$

*5 How many moles of each solute are contained in the following solutions? (Express your answer as a decimal, if necessary.)

a 250 cm^3 of potassium dichromate(VI), 0.100 mol dm^{-3}
b 25 cm^3 of sodium chloride, 0.100 mol dm^{-3}
c 10 cm^3 of sodium chloride, 2.00 mol dm^{-3}
d 12.2 cm^3 of nitric acid, 1.56 mol dm^{-3}
e 12.2 cm^3 of sulphuric acid, 1.56 mol dm^{-3}

*6 What is the concentration, in $mol\,dm^{-3}$, of each of the following solutions?

a $5.85\,g$ of sodium chloride, $NaCl$, in $1000\,cm^3$ of solution
b $5.85\,g$ of sodium chloride, $NaCl$, in $250\,cm^3$ of solution
c $3.16\,g$ of potassium manganate(VII) $(KMnO_4)$ in $2\,dm^3$ of solution
d $6.20\,g$ of sodium thiosulphate $(Na_2S_2O_3 \cdot 5H_2O)$ in $250\,cm^3$ of solution
e $5.62\,g$ of hydrated copper(II) sulphate $(CuSO_4 \cdot 5H_2O)$ in $250\,cm^3$ of solution

*7 Calculate the concentration, in $mol\,dm^{-3}$, of:

a Ethanol (C_2H_6O), $23\,g$ in $1\,dm^3$ of solution
b Hydrogen ion in $1\,dm^3$ of solution containing $3.65\,g$ of hydrogen chloride (assume that the hydrogen chloride is fully ionized)
c Hydroxide ion in $1\,dm^3$ of solution containing $17.1\,g$ of barium hydroxide, $Ba(OH)_2$ (assume that the barium hydroxide is fully ionized)
d Sulphate ion in a solution of aluminium sulphate, $Al_2(SO_4)_3 \cdot 12H_2O$, of concentration $0.1\,mol\,dm^{-3}$
e Aluminium ion in a solution of aluminium sulphate, $Al_2(SO_4)_3 \cdot 12H_2O$, of concentration $0.1\,mol\,dm^{-3}$

8 Figure 1.9 is a bar chart of the atomic volume of some elements arranged in order of their atomic number at 298 K and 1 atmosphere pressure. The actual atomic numbers have been omitted. Each letter on the horizontal axis represents an element, but the letter is not the chemical symbol for the element. Each element differs from the next by one atomic number. (The volume of 1 mole of atoms of element J is $24.3\,dm^3$ at 298 K and 1 atmosphere pressure.)

a Which of these elements would you expect to be a gas at room temperature and pressure?
b Which of these elements would you expect to be an alkali metal?
c Which of these elements will contain the greatest number of atoms per cubic centimetre of the element?
d The density of element G is $5.73\,g\,cm^{-3}$; calculate the mass of 1 mole of atoms of the element.
e Is it possible, from the information given in the figure, to determine which of these elements has the highest density in grams per cubic centimetre? If so, state which one has the highest density and give your reasons. If not, explain why not.

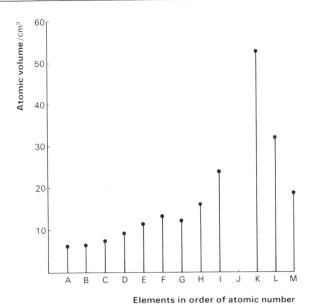

Figure 1.9

9 In a certain electrolysis experiment using silver nitrate solution as electrolyte, 0.216 g of silver was deposited by a steady current of 0.200 ampere flowing for 960 seconds. The relative atomic mass of silver is 108. The charge on the electron is 1.60×10^{-19} coulomb. What value for the Avogadro constant do these figures give?

10 The number of alpha particles emitted from a sample of radium was measured by a Geiger counter and found to be 8.20×10^{10} second^{-1}. The same sample produced 0.0790 cm^3 of helium in 300 days (at s.t.p.). What value do these figures give for the Avogadro constant?

The Periodic Table 2: the elements of Groups I and II

Group I		**Group II**	
Lithium	Li	Beryllium	Be
Sodium	Na	Magnesium	Mg
Potassium	K	Calcium	Ca
Rubidium	Rb	Strontium	Sr
Caesium	Cs	Barium	Ba
Francium	Fr	Radium	Ra

In this Topic we shall consider the properties of the elements of Groups I and II. During this investigation, attention will be directed to two main features:

1 The similarities and differences which exist between the elements of Groups I and II of the Periodic Table.

2 The trends in properties to be found in Groups I and II.

The elements of these two groups are predominantly metals. Only beryllium, the details of whose chemistry we will not be considering, shows any significant non-metallic character.

Figure 2.1
The elements of Groups I (below) and II (opposite).   *Photographs, Jit Baran.*

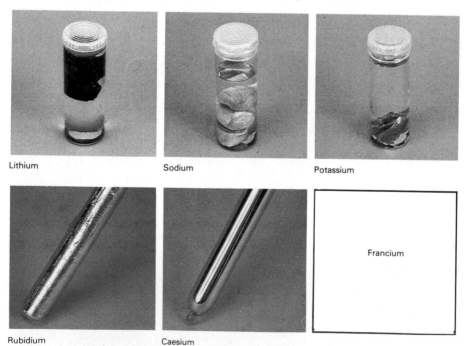

Lithium   Sodium   Potassium

Rubidium   Caesium   Francium

2.1
THE ELEMENTS OF GROUPS I AND II

In this section you will be gathering some information about the elements of Groups I and II, and then trying to find any trends that might exist in their properties. The information is best recorded in your notebook in a series of tables, as now suggested. Questions are given after each table; you should write the answers to these questions in your notebook after the table, in such a way as to make clear what the questions were.

TABLE A
The physical appearance of the elements

Draw up a table with the headings shown overleaf. Describe the appearance of each element, using samples of the elements where possible, or the photographs in figure 2.1. Look up the melting points and boiling points in the *Book of data* and record these. Note that the *Book of data* gives these figures in kelvins; you might record them in °C in your table by subtracting 273 from the kelvin value. This is easily done using a calculator, especially if it has a constant button. Record your chosen temperature units at the heads of the columns. Densities can also be found in the *Book of data*, in the same table as the melting and boiling points.

Beryllium

Magnesium

Calcium

Strontium

Barium

Radium

Element	Atomic radius/nm	Ion formed	Ionic radius/nm
Lithium	0.157	Li^+	0.074

Table B

Questions

1 What generalization can be made about the size of the ion of one of these metals, relative to the size of its atom?

2 What are the trends in atomic and ionic radii with increasing atomic number in each group?

TABLE C
The reactions of the metals with oxygen and water

You may be shown the reactions of some of these metals with oxygen and with water, or you may obtain a description of these reactions from a textbook of inorganic chemistry. Record the observations that you make, and write equations for the chemical reactions involved.

Note. Sodium reacts with *water* according to the equation

$$2Na(s) + 2H_2O(l) \longrightarrow 2NaOH(aq) + H_2(g)$$

and the other alkali metals react similarly. The equation for the reaction of calcium with water is

$$Ca(s) + 2H_2O(l) \longrightarrow Ca(OH)_2(aq) + H_2(g)$$

and this is typical of the alkaline earth metals.

The reactions with *oxygen* are not so regular. Sodium, for example, forms mainly sodium peroxide

$$2Na(s) + O_2(g) \longrightarrow Na_2O_2(s)$$

whereas other alkali metals form different oxides.

Magnesium gives magnesium oxide

$$2Mg(s) + O_2(g) \longrightarrow 2MgO(s)$$

and the other Group II metals react similarly.

Element	Reaction with water	Reaction with oxygen

Table C

Question

What trends are noticeable in the vigour of the reactions?

TABLE D
Some compounds of the elements of Groups I and II

Examine the chlorides, sulphates, and nitrates of the elements and record in a copy of this table their appearance and chemical formulae, including any water of crystallization.

Formulae may be printed on the labels of the bottles containing the various compounds, or they may be found from the table of properties of inorganic compounds in the *Book of data*.

Element	Chloride	Sulphate	Nitrate

Table D

Question

What connection is there between the tendency of these compounds to be hydrated and the positions of the metals in the Groups?

2.2
THE OXIDES AND HYDROXIDES OF GROUPS I AND II

WARNING: Some of the oxides and hydroxides of the metals suggested for use in these experiments are fine powders which are very harmful to the eyes. Their mixtures with water are also harmful. You *MUST* wear safety glasses.

Oxides of metals are normally basic, and so they react with acids to give salts. If basic oxides are mixed with water, it is possible that they may react to form hydroxides, and if the hydroxides are soluble, the solution will be alkaline.

EXPERIMENT 2.2
An investigation of the action of water on some Group I and Group II metal oxides

Procedure

To 0.1 g of sodium peroxide, calcium oxide, and magnesium oxide add about 5 cm^3 of pure water. Find the pH of the mixtures, using Full-range Indicator paper. Then add 1M hydrochloric acid until the mixture is just acidic. Enter the results in a table like the one that follows.

Name of oxide	Formula	pH of mixture with water	Observations on adding 1M hydrochloric acid

Questions

Here are some questions about the results of the experiments. Write the answers into your notebook in such a way as to make it clear what each question was.

1 What signs were there that reactions occurred when water was added to the oxides?

2 Bearing in mind the second paragraph of this section, what is the evidence that the oxides are basic?

3 Since the mixtures of the oxides with water were alkaline, hydroxides were presumably formed. Write equations for the reactions of the oxides with water. It should be noted that sodium peroxide, Na_2O_2, is not the typical oxide of sodium, which would have the formula Na_2O. Sodium peroxide reacts with water to give not only the hydroxide but also oxygen.

4 Write equations for the reactions of the hydroxides with hydrochloric acid.

5 What can be said about the solubility in water of the chlorides of each of the metals whose oxides you used?

6 What differences, if any, would you expect to find if you had used dilute sulphuric acid or dilute nitric acid in the experiment?
(Use the table of properties of inorganic compounds in the *Book of data* to find the solubilities of the sulphates and nitrates of the metals.)
You will be able to check your predictions experimentally, in section 2.5.

2.3
THE SOLUBILITY OF CALCIUM HYDROXIDE

In the last section you will have noticed that calcium hydroxide is soluble in water, but not very soluble. In this section you will do an experiment to find out just how soluble it is.

EXPERIMENT 2.3
To find the solubility of calcium hydroxide in water by titration

Procedure

1 Put about $100\,cm^3$ of pure water in a conical flask and add one spatula measure of solid calcium hydroxide. Fit the flask with a cork or rubber stopper and agitate the mixture thoroughly. Allow the mixture to stand for at least 24 hours so that the water becomes saturated with the calcium hydroxide.

2 Carefully decant the solution into a filter funnel and filter paper over a second conical flask, so as to collect the saturated solution of calcium hydroxide.

3 Titrate $10\,cm^3$ portions of this solution with 0.05M hydrochloric acid, using methyl orange or bromophenol blue as indicator. Repeat the titrations until two successive results agree to within $0.1\,cm^3$. Your teacher will show you, if necessary, how to use the burette and pipette properly.

lutetium in the Periodic Table) are so similar chemically that the only methods available for their separation before the introduction of ion exchange resins were based on repeated, tedious fractional crystallization of their salts. With suitable ion exchange procedures it is possible to make use of small but significant changes in the affinity of a resin for the different elements, so that all fourteen can be cleanly separated. These methods were worked out in America during World War II, and have since been developed on a commercial scale. We can now buy kilogram quantities of pure, individual elements and their compounds.

Similar procedures are used to separate the highly radioactive man-made elements which follow uranium in the Periodic Table; it is questionable whether many of these could have been isolated by any other means, since the quantities formed are so minute and they exist for such a short time that a very swift, clean, and simple method is needed to separate them.

Regeneration

Ion exchange resins repeatedly used are eventually unable to exchange any more ions. When this happens the resins may be 'regenerated' by soaking them in a high concentration of the ions which were originally present; this reverses the ion exchange process.

2.5
TRENDS IN SOLUBILITY OF GROUP II METAL SALTS

In this section we shall examine some reactions between the cations of Group II elements and various anions in solution and look for any trends that can be seen in the solubilities of the compounds that may be formed.

A quick check of the aqueous solutions available in your laboratory, or a look at values for solubility in the table of inorganic compounds in the *Book of data*, should enable you to state whether or not the salts of the alkali metals are usually readily soluble in water.

You can find out which salts of Group II elements are soluble in water by doing experiment 2.5, and at the same time check on the predictions that you made earlier (experiment 2.2).

EXPERIMENT 2.5
To investigate the solubility of some salts of Group II metals

For the investigation of the solubility of the salts of the Group II metals, you will need aqueous solutions which are approximately 0.2M with respect to their cations. Since all nitrates are soluble they are convenient salts to use.

Also needed are solutions of sodium salts which are approximately 0.2M with respect to their anions.

The range of solutions required is listed in table 2.2.

Group II cations 0.2M (aq)	Anions having a single charge 0.2M (aq)	Anions having two charges 0.2M (aq)
Mg^{2+}	Cl^-	CO_3^{2-}
Ca^{2+}	Br^-	SO_4^{2-}
Sr^{2+}	OH^-	CrO_4^{2-}
Ba^{2+}	NO_3^-	$C_2O_4^{2-}$

Table 2.2
Solutions required for experiment 2.5.

The experimental work is best carried out in groups of four students so that all four cations can be studied simultaneously.

WARNING: barium salts (except the sulphate) are poisonous, as are ethanedioates (oxalates).

Procedure

Mix portions of a cation solution with the various anions in turn and leave the results in your test-tube rack. Decide whether the cation forms salts with singly charged ions that are generally soluble, generally insoluble, or neither. For example, if the four anions give a result of three precipitates and one solution the result can be classed as 'generally insoluble', while two precipitates and two solutions should be classed as 'neither'.

Make the same decisions about the behaviour of the cation with the anion of charge -2. Exchange test-tube racks with other students in order to complete your results for the remaining three cations.

Devise a suitable table in which to record your generalizations, keeping one column in which to note any anions which were exceptions to the general rule.

Using your table as a guide, consider whether further generalization is possible about the Group II metals. Is any one cation rather different in behaviour from the others? Does any particular anion feature in the exceptions column?

The study of solubility may now be taken a little further by looking for trends in solubility down the Group II metal group. To do this, select any one anion which gives insoluble salts with the Group II metals and use it for the following test.

Put equal portions of solutions of the four cations in separate test-tubes and add just one drop of the selected anion solution to each.

Can you observe any difference in the amounts of precipitate formed? Add a second and then a third drop if necessary in order to make a decision. For the selected anion, does solubility increase, decrease, or not appear to change on going down Group II?

Use the *Book of data* to confirm your qualitative observation and note the trend in solubility for other salts, both soluble and almost insoluble. Plotting a rough graph may help to make any general trends stand out. Are there general trends, and if so, are any salts clear exceptions?

2.6
THE EFFECT OF HEAT ON THE CARBONATES AND NITRATES

We shall now examine the effect of heat on the carbonates and nitrates of the elements of Groups I and II, and look for any trends that can be seen in the ease with which they decompose.

EXPERIMENT 2.6a
To investigate the effect of heat on the carbonates of the elements of Groups I and II

When carbonates are heated, they may decompose, and if they do, carbon dioxide is evolved and the oxide of the metal remains. Copper carbonate, for example, decomposes according to the equation

$$CuCO_3(s) \longrightarrow CuO(s) + CO_2(g)$$

Devise and use a technique for estimating the comparative readiness with which the carbonates of the Groups I and II metals decompose.

Record your results in the form of a table:

Group I	Effect	Ionic radius of cation	Group II	Effect	Ionic radius of cation
Li_2CO_3			$MgCO_3$		
Na_2CO_3			$CaCO_3$		
K_2CO_3			$SrCO_3$		
			$BaCO_3$		

There is a connection between the trend of behaviour down each group and the ionic radius of the cation. Discuss with your teacher the reasons for the connection and record in your notebook the outcome of your discussion.

Write equations for any decompositions that take place.

EXPERIMENT 2.6b
To investigate the effect of heat on the nitrates of the elements of Groups I and II

When nitrates decompose, one of two things can happen:

1 Oxygen is evolved and the nitrite of the metal (containing the ion NO_2^-) is formed.

2 Oxygen and the brown gas nitrogen dioxide are evolved and the oxide of the metal remains.

Lead nitrate, for example, decomposes according to the equation

$$2Pb(NO_3)_2(s) \longrightarrow 2PbO(s) + 4NO_2(g) + O_2(g)$$

Either investigate practically the effect of heat on nitrates (use a fume cupboard because nitrogen dioxide is poisonous) or look up the results in a textbook of inorganic chemistry.

Draw up a similar table to the one that you made for the carbonates and once again compare the results with the trends in the ionic radii of the metal ions.

Write equations for any decompositions that may take place.

2.7
FLAME COLOURS

Many elements give characteristic colours when their compounds are placed in a Bunsen burner flame. These colours are caused by electrons losing the energy that they have gained from the heat of the Bunsen burner flame. They will be discussed in Topic 4. In the meantime, experiment 2.7 will enable you to gather some information about these colours.

EXPERIMENT 2.7
An investigation of the flame colours of the elements of Groups I and II

Although all compounds of a particular element give the same flame colour, chlorides are the most satisfactory to use because they are usually volatile at temperatures attainable in the Bunsen burner flame.

Clean a platinum or nichrome wire by heating it in a non-luminous Bunsen flame, dipping it into a little concentrated hydrochloric acid (in a crucible or small watch glass), and heating it again. Continue this until the wire imparts no colour to the flame.

Pour the impure acid away and take a fresh portion. Dip the clean wire into the acid and then into a small portion of powdered compound on a watch glass. Use chlorides where possible, otherwise nitrates or carbonates. Hold the wire so that the powdered solid is in the edge of the flame and note any coloured flame which results. The colour disappears fairly quickly but can be renewed by dipping the wire into acid again and reheating. Observe the flame through a diffraction grating or direct vision spectroscope and identify as many coloured lines as you can. You may need to take a fresh sample of solid for this and to do the experiment in a darkened corner of the laboratory. The wire must be cleaned before examining a new compound, and a fresh portion of acid will be needed for this.

Record in your notebook the coloured lines that you see for each element that you examine.

BACKGROUND READING 2
The role of calcium and magnesium in agriculture

An adequate supply of calcium compounds in the soil is essential for healthy plant growth. Calcium itself is an essential constituent of plants, and it is also important for another reason. Calcium compounds are the principal factor in controlling the pH of the soil, and this affects the ability of plants to absorb nutrients through the roots.

The pH of the soil influences the concentration of plant nutrients in the soil solution and hence their availability. For example at a pH of about 5 the concentration of aluminium and manganese is higher than at a pH of 7. Some plants grow best at a low soil pH and are checked at higher values. Tea is a well known example of a crop which thrives in very acid soils and it contains far more aluminium than most plants.

Some species of forest trees do not thrive in soils of high pH. Sitka Spruce, for example, was found to make the best growth at pH 5 and failed to grow well on neutral and alkaline soils. But growth was depressed below pH 5 – a narrow range of optimum pH.

On the other hand, sugar beet does not grow well under acid conditions and the optimum pH for this crop is around 6.5–7.0.

Crops are roughly graded in their tolerance for soil acidity: lucerne, sugar beet, and barley are only considered suitable for neutral or slightly acid soils (pH 6.5–7.0); wheat grows well on more acid soils (pH 6.0–6.5); and potatoes and rye on soils of pH 5.0 – too acid for sugar beet and barley.

Figure 2.4 shows in the foreground two barley plots. The one on the right contains soil with low calcium content; it has become acid in reaction and the barley crop has failed there. The plot on the left has been well limed and has a good calcium content and a pH of about 6.5; the barley crop is good there.

Figure 2.4
Barley on a limed plot (*left*) and an unlimed plot (*right*).
Photograph, Rothamsted Experimental Station.

Calcium in the soil is mostly in the exchangeable form (see below), with some reserve as calcium carbonate. The soil is made up of both inorganic and organic materials. An important part consists of clay minerals which carry a negative charge; the humus derived from the organic matter is similarly charged. These charges are balanced by cations in the soil, mainly the H^+, Ca^{2+}, Mg^{2+}, K^+, and Na^+ ions. If a soil is washed with a solution of, say, calcium chloride of concentration 1 mol dm^{-3}, the washings have a lower concentration of calcium ions than the initial solution, but contain varying amounts of the cations just mentioned. Ions replaced in this way are referred to as 'exchangeable'; the clay and humus content of the soil acts as an ion exchange resin would.

Rain water, which contains carbonic acid, H_2CO_3, leads, as it percolates through the soil, to the replacement of cations such as Ca^{2+} by H^+. Fertilizers such as ammonium sulphate $(NH_4)_2SO_4$, are also involved in cation exchange.

$$Ca\text{-soil} + 2NH_4^+(aq) \longrightarrow (NH_4)_2\text{-soil} + Ca^{2+}(aq)$$

a What indicator would you use for the titration and how does its colour change?

b What difficulty might be encountered while doing the titration?

c Write an equation to represent the reaction occurring on the ion exchange resin.

d Use the data to calculate the solubility of calcium chromate in moles per cubic decimetre.

*6 Using a supplier's catalogue for prices, find the ratio of the price of calcium to the price of magnesium

a per gram,

b per mole of atoms,

c per cm^3.

d Calcium compounds are nearly twice as abundant as magnesium compounds, but calcium is more expensive than magnesium. Make a comprehensive list of possible reasons for this anomaly.

*7 Suppose you wanted to make as much pure calcium ethanedioate (oxalate, CaC_2O_4) as possible, given a solution of 1.9 g of sodium ethanedioate in 50 g of water (20 °C) and a supply of 0.1M calcium chloride solution.

a Calculate the minimum volume of calcium chloride solution which you would require. Give an equation for the reaction.

b Describe the procedure.

c What is the maximum mass of calcium ethanedioate you could expect to obtain?

d Give all the reasons why you are unlikely to obtain in practice the mass given in answer to **c**.

*8 Give an account of the general trends in

a the solubility and

b the effect of heat on

the hydroxides, carbonates, and sulphates, of sodium, magnesium, calcium, strontium, and barium.

*9 Make a general comparison of the melting points and the enthalpy changes of fusion (per mole of atoms) of the following elements: sodium, magnesium, calcium, strontium, barium, phosphorus (white), sulphur, chlorine, bromine, and iodine.

TOPIC 3
The properties of gases

In this Topic we shall review some of the properties of gases and try to understand why they have these properties, using a molecular picture of their behaviour.

We shall also meet some experimental methods for finding the relative masses of molecules and atoms, based on the properties of gases: methods that were important landmarks in the historical development of chemistry.

3.1
GAY-LUSSAC'S LAW

Having studied the results of a large number of reactions involving gases, the French scientist Gay-Lussac, in the years just after 1800, saw that the volume measurements were very simply related. For example, when hydrogen and oxygen combine

2 volumes of hydrogen + 1 volume of oxygen → 2 volumes of steam

and for the hydrogen and chlorine reaction,

1 volume of hydrogen + 1 volume of chlorine
→ 2 volumes of hydrogen chloride

He expressed this in his *Law of Gaseous Combining Volumes*, which he published in 1809:

' When gases react, the volumes in which they do so bear a simple ratio to one another, and to the volume of the products if gaseous, all volumes being measured under the same conditions of temperature and pressure.'

Avogadro's theory

An explanation of the Law in terms of atoms and molecules was sought, and in 1811 the Italian scientist Avogadro put forward his theory:

' Equal volumes of gases, under the same conditions of temperature and pressure, contain equal numbers of molecules.'

Boyle's Law can be expressed in symbols as $V_0 \propto \dfrac{1}{p}$

This can be written as $V_0 = k\dfrac{1}{p}$, or $pV_0 = k$, where k is a constant.

Charles's Law can be expressed as $V_0 \propto T$.

This can be written as $V_0 = k'T$, or $\dfrac{V_0}{T} = k'$, where k' is another constant.

Combining these two relations we have $\dfrac{pV_0}{T} = R$

or, multiplying both sides by T

$$pV_0 = RT$$

where R is a constant known as the *gas constant*.

For n moles of gas, volume V, the equation becomes

$$pV = nRT$$

and is known as the *Ideal Gas Equation*.

If p is measured in atmospheres, V in cubic decimetres, and T in kelvins, the gas constant R has units of $atm\,dm^3\,K^{-1}\,mol^{-1}$. The numerical value of R in these units is 0.082.

The most convenient procedure when calculating the results of experimental work is to 'correct' observations to standard temperature and pressure (s.t.p., 273 K or 0 °C and 1 atm or 760 mmHg*), using the Gas Laws, in order to obtain a proper comparison between one set of results and another.

Correction to s.t.p. – a simple example

The volume of a certain mass of gas was measured at 40 °C and 750 mmHg pressure and found to be 65 cm³. What is its volume at s.t.p.?

---

*The internationally accepted unit of pressure is the pascal, Pa. One pascal is defined as a pressure of one newton per square metre, $N\,m^{-2}$.

For many practical purposes, gas pressures are normally recorded as a height of a column of mercury in a barometer, and one 'atmosphere' is the pressure which will support a column of mercury 760 mm high. It is related to the international unit in the following way: 1 atmosphere = 101 325 Pa.

In this book we shall normally be concerned with the comparison of gas pressures, and will use the practical unit of measurement, the mmHg.

At 313 K and 750 mmHg, volume of gas is 65 cm^3

at 273 K and 760 mmHg, volume is $65 \times \dfrac{273}{313} \times \dfrac{750}{760}$ cm^3

i.e. volume of gas at s.t.p. is 55.9 cm^3

Notice that if the pressure on the gas is increased its volume decreases, *i.e.* is multiplied by $\dfrac{750}{760}$, not by $\dfrac{760}{750}$.

Correction to s.t.p. can also be made by direct substitution in the Ideal Gas Equation.

At s.t.p., one mole of molecules of hydrogen, H_2 (that is 2 g of hydrogen), occupies 22.4 cubic decimetres. By Avogadro's theory it therefore follows that 22.4 cubic decimetres (at s.t.p.) of any other gas contains the same number of molecules as one mole of hydrogen molecules. In other words, one mole of molecules of any gas occupies 22.4 cubic decimetres at s.t.p.

3.4
FINDING THE RELATIVE MASSES OF MOLECULES

This section describes some experiments you can do to find the relative masses of the molecules of some gases and volatile liquids.

EXPERIMENT 3.4a
To find the relative masses of molecules of gases

In this experiment we shall find the mass of a stoppered flask full of the gas under investigation, and then find its volume. We shall then calculate the mass of the same volume of hydrogen and, from these results, work out first the vapour density and then the relative molecular mass of the gas.

Suitable gases to use include oxygen, nitrogen, methane, carbon dioxide, and hydrogen chloride.

Oxygen and *nitrogen* are most easily obtained from cylinders of the gases. *WARNING:* the gas in these cylinders can be at a pressure of up to 100 atmospheres. Its sudden release can cause considerable damage.

The domestic gas supply in most areas of Britain is a convenient source of *methane*.

Carbon dioxide can also be obtained from a cylinder. If it is made by the action of dilute hydrochloric acid on marble chips, it must be dried by passing it through a wash-bottle containing concentrated sulphuric acid before it is used.

Hydrogen chloride is made by the action of concentrated sulphuric acid on sodium chloride. If it is used, the flasks must be filled in a fume cupboard.

All masses should be found on a balance capable of weighing to 3 decimal places of grams.

Masses, and other readings, should be recorded in a table in your notebook, which can be prepared before you start the experiment. The following entries will be needed:

Mass of flask and air	g
Mass of flask and ...... gas	g
Volume of flask	cm^3
Laboratory temperature	°C
Atmospheric pressure	mmHg

Procedure

1 Weigh accurately a stoppered flask of approximately $100 \, cm^3$ capacity. It will, of course, be filled with air.

2 Remove the stopper and insert a glass delivery tube connected to a suitable gas generator into the flask so that the open end of the tube is very near the bottom of the flask. Pass gas into the flask until you judge that all the air is displaced. This should take 1–2 minutes. Remove the delivery tube slowly and stopper the flask immediately. Why should the delivery tube be removed slowly?

3 Weigh the flask plus gas. To check that all the air was displaced, pass gas through the flask again for 1–2 minutes and reweigh. If the two masses agree to within 0.001 g, it can be assumed that little or no air was in the flask when the first mass of flask plus gas was found. If the masses do not agree repeat the process until agreement between successive weighings is obtained.

4 To find the capacity of the flask, mark the position of the bottom of the stopper on the neck of the flask (use a grease pencil or felt-tipped marker). Fill the flask with water to the level of this mark and find the volume of water added by pouring it into a measuring cylinder. A volumetric flask is convenient for this experiment. If a flask of this type is used, its capacity can be found more accurately by filling to the graduation mark with water and then running in more water from a burette until the level reaches the mark indicating the bottom of the stopper.

5 Record the laboratory temperature (°C) and atmospheric pressure (mmHg).

Calculations

1 Find the volume which the air (and the gas used) would have at s.t.p.

2 Taking the density of dry air at s.t.p. as $1.293 \, g \, dm^{-3}$, calculate the mass of air in the vessel used.

3 Subtract the mass of air found in **2** from the mass of the flask full of air, in order to obtain the mass of the flask empty.

4 Subtract the mass of the flask empty from the mass of the flask full of gas, in order to obtain the mass of the gas taken.

5 Taking the density of hydrogen at s.t.p. as $0.090\,g\,dm^{-3}$, calculate the mass of this volume of hydrogen.

6 Divide the mass of the gas found in **4** by the mass of hydrogen found in **5**, to obtain the vapour density of the gas.

7 Multiply the vapour density by 2 to obtain the relative molecular mass of the gas.

Additional exercise

Estimate the accuracy which you could attach to your result. How do you think the method you used could be improved to give more accurate results?

EXPERIMENT 3.4b
To determine the relative molecular mass of molecules of a liquid of fairly low boiling point

By finding the volume of the vapour formed from a known mass of liquid, at a known temperature and pressure, it is possible to calculate the relative molecular mass of the liquid. The simplest way to do this is to allow a known mass of liquid to vaporize in a heated graduated syringe, the temperature of which is at least 20 °C above the boiling point of the liquid. If steam is used to heat the syringe, as in the method described below, liquids of boiling point lower than 80 °C must be used. A glass syringe is essential for this experiment; plastic syringes soften at steam temperature.

The apparatus used is shown in figure 3.1. A metal can (or a glass flask) fitted with an outlet tube and a safety tube can be used as a steam generator. Put a beaker under the outlet from the steam jacket to catch drops of water from the condensing steam.

Procedure

1 Assemble the apparatus without the self-sealing rubber cap and the hypodermic syringe. Draw about $5\,cm^3$ of air into the large graduated syringe and fit the self-sealing cap over its nozzle. Pass steam through the jacket until the thermometer reading and the volume of air in the syringe reach steady values. Record the temperature and air volume. Continue to pass steam through the jacket while you prepare the hypodermic syringe.

2 Fit a hollow needle to the hypodermic syringe. Push in the plunger, put the needle into the liquid that is to be investigated, and draw about $1\,cm^3$ of

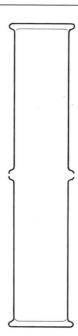

Figure 3.2

Thomas Graham, in the years around 1830, studied the rates at which different gases diffused, and summarized his findings in his Law of Gaseous Diffusion.

'The rate of diffusion of a gas is inversely proportional to the square root of its density.'

If we want to compare the rates of diffusion of two gases (at the same temperature and pressure), **Graham's Law** tells us that

$$\frac{\text{rate of diffusion of gas A}}{\text{rate of diffusion of gas B}} = \sqrt{\frac{\text{density of gas B}}{\text{density of gas A}}}$$

If gas A is hydrogen, then we have

$$\frac{\text{rate of diffusion of hydrogen}}{\text{rate of diffusion of gas B}} = \sqrt{\frac{\text{density of gas B}}{\text{density of hydrogen}}}$$

Now

$$\frac{\text{density of gas B}}{\text{density of hydrogen}} = \frac{\text{mass of unit volume of gas B}}{\text{mass of same volume of hydrogen}}$$

$$= \text{vapour density of gas B}$$

$$= \tfrac{1}{2} \times \text{relative molecular mass of gas B}$$

So

$$\frac{\text{rate of diffusion of hydrogen}}{\text{rate of diffusion of gas B}} = \sqrt{(\tfrac{1}{2} \times \text{relative molecular mass of gas B})}$$

or

$$\text{relative molecular mass of gas B} = 2 \times \left(\frac{\text{rate of diffusion of hydrogen}}{\text{rate of diffusion of gas B}}\right)^2$$

A comparison of rates of diffusion therefore gives us another way of finding the relative molecular mass of a gas.

You may be shown an experiment to illustrate the different rates of diffusion of different gases.

3.6
THE KINETIC THEORY

We have met three Gas Laws in this Topic that we can profitably consider a little further at this stage: Boyle's Law, Charles's Law, and Graham's Law. If we make certain assumptions about gas molecules, we can see that these laws describe behaviour that is to be expected of gases. The assumptions, and the arguments which now follow, are known as the *kinetic theory of gases*.

Let us suppose that we have one mole of molecules of a gas, consisting of the Avogadro constant L molecules, held in a cubic container of volume V_0 cm^3. The molecules will be moving at random in all directions, but their movement can be resolved in three directions at right angles to each other. So we may imagine $\tfrac{1}{6}$ of the molecules moving to the left and $\tfrac{1}{6}$ to the right; $\tfrac{1}{6}$ up and $\tfrac{1}{6}$ down; $\tfrac{1}{6}$ forward and $\tfrac{1}{6}$ back. We shall assume that the molecules are extremely small compared with the space in which they can move, and that they have no effect on one another.

We will start by working out an expression for the *pressure* exerted by the gas on the walls of the container. Now pressure is *force* per unit area, so so

$$\rho = \frac{mL}{V_0} = \frac{3p}{u^2}$$

Rearranging,

$$u = \sqrt{\frac{3p}{\rho}}$$

So for a given pressure,

$$u \propto \sqrt{\frac{1}{\rho}}$$

The rate of diffusion is clearly related directly to the speed of the molecules, so

$$\text{rate of diffusion} = k'' \sqrt{\frac{1}{\rho}} \quad \textbf{which is Graham's Law.}$$

One more interesting point. Earlier in the Topic we saw that Boyle's Law and Charles's Law could be combined in one expression, namely

$$pV_0 = RT$$

Returning to the expression

$$pV_0 = \frac{1}{3}Lmu^2$$

$$= \frac{2}{3}L(\tfrac{1}{2}mu^2)$$

Now $\frac{1}{2}mu^2$ is the average kinetic energy of one molecule of the gas, and this is proportional to the temperature T of the gas.

$$T \propto \frac{1}{2}mu^2$$

So we can write

$$pV_0 \propto \frac{2}{3}LT$$

Incorporating the $\frac{2}{3}$ into the constant of proportionality, we have

$$pV_0 = LkT$$

Now we have a new meaning for R, the gas constant; it is the Avogadro constant multiplied by a factor k. This factor is known as the **Boltzmann constant**; it is, in a sense, the gas constant for a single molecule, and we shall meet it again in later Topics. It has the value of $1.38 \times 10^{-23} \, \text{J K}^{-1}$.

We shall write the Ideal Gas Equation as

$$pV = nLkT$$

whenever it is referred to later in this book.

3.7
THE CHANCE BEHAVIOUR OF MOLECULES

Before we leave the topic of gases we shall try to take this molecular picture of their behaviour a stage further. The content of this section may seem rather difficult at first, but do not worry about it; we shall be returning to these ideas several times during the course, and with each reading will come a greater understanding. If you can master the ideas which this section introduces, and build on them with the aid of sections that appear in several later Topics in this book (4, 6, and 10), it will greatly add to your understanding of the behaviour of matter.

The kinetic theory of gases might be said to be the working of the consequences of one simple principle, namely

molecules don't care.

The pressure of a gas is decided simply by the fact that the molecules move about *at random*, hitting the walls of a container when they happen to meet them. Gases do not exert a pressure because the molecules 'want to escape', or because they 'feel confined'. Molecules cannot 'know' or 'like' anything, let alone the Gas Laws. They just move at random, and the Gas Laws result.

In this section, we shall start to explore some ideas, with the eventual aim of finding an answer to the question, 'Why do chemical reactions happen?'. We shall begin by looking further into the principle

molecules don't care

We shall see how this principle helps to explain which way a reaction will go: why at room temperature H_2 molecules do not dissociate, why hydrogen and oxygen do combine to make water, why alcohol distils into a vapour at high temperatures but condenses at low temperatures, why high pressures help the combination of hydrogen and nitrogen to make fertilizers, or the linking make 10^{30} moves in 10^{24} seconds. Now the Universe is about 10^{10} years old, which is about 3×10^{17} seconds. So even a fast computer would need a lot of Universe-lifetimes to make it reasonable to see all 100 in one half.

Now let's try it for one mole of gas, with 6×10^{23} molecules. The total number of arrangements that are possible when two halves become accessible to this number of molecules is

$$2^{(6 \times 10^{23})}$$

If you try to work this out on your calculator, it won't be able to do it! The number is too big. But we can work it out if we use logarithms.

We know that

$$\log (2 \times 2 \times 2) = \log 2 + \log 2 + \log 2 = 3 \log 2$$
or
$$\log 2^3 = 3 \log 2$$

So

$$
\begin{aligned}
\log 2^{(6 \times 10^{23})} &= (6 \times 10^{23}) \log 2 \\
&= (6 \times 10^{23}) \times 0.3 \quad \text{(using logarithms to base 10)} \\
&\approx 2 \times 10^{23} \qquad\qquad \text{roughly}
\end{aligned}
$$

Now this is the *log* of the number we want, not the number itself. But just as the log of 1 000 000 is 6, so the log to base ten of any number is the number of noughts in it. So the big number we want is 1 followed by 2×10^{23} noughts. Even if we wrote it in letters the size of atoms, 10^{-10} m across, it would occupy 2×10^{13} m, which is more than 100 times the distance from the Sun to the Earth. (If we used letters 1 mm across, the number would stretch to a distant star.)

We reach an important conclusion: the reason why gases diffuse into empty spaces is just that that is *overwhelmingly likely by chance alone*.

So we can compute the factor by which the number of arrangements of N particles is multiplied, when we double the volume available to them; it is 2^N.

It is **2 to the power** N (not, say, $2N$) because *numbers of ways multiply*. A good way to deal with things that multiply, to give huge numbers, is to use logarithms, so making things *add*. That is why, if you look up a book of physical chemistry, you will find quite a number of equations which use logarithms. So you have to get used to them!

As will be seen later, it is better in most chemical situations to use natural logarithms (logarithms to base e, e = 2.718 ...) rather than logarithms to base 10. Natural logarithms are obtained by using the 'ln' button on a calculator, and we shall use them in all future work in this book.

Example Suppose a mole of neon atoms doubles its volume. Let the number of arrangements of neon atoms before doubling the volume be W_1, and the number of arrangements accessible to the atoms after doubling the volume be W_2. Then

$$W_2/W_1 = 2^{6 \times 10^{23}}$$
and
$$\ln(W_2/W_1) = 6 \times 10^{23} \ln 2$$
$$\approx 4 \times 10^{23}$$

This means that on doubling the volume, the number of possible arrangements of 1 mole of neon atoms is multiplied by something (2.718 ...) **to the power** of 4×10^{23}. Even when working in logarithms, we still have very large numbers. To enable us to handle these large numbers more easily, it is usual to scale them down to make them more reasonable.

Boltzmann's constant $k = 1.38 \times 10^{-23} \, \mathrm{J\,K^{-1}}$ is the usual choice of scale constant. This is chosen because it happens that quantities concerned with numbers of ways are involved in the relation between energy and temperature. If we choose a scale constant to have this size and the units $\mathrm{J\,K^{-1}}$, then temperatures stay on the kelvin scale.

So this gives the effect of doubling the volume of 1 mole of neon atoms as

$$k \ln(W_2/W_1) = 1.38 \times 10^{-23} \times 4 \times 10^{23} = 6 \, \mathrm{J\,K^{-1}}$$

Entropy

Quantities of the kind

$k \times$ logarithm of factor by which number of ways is multiplied

will turn up again and again, when we consider arrangements of molecules and their energy in chemical reactions. So they have a special name. They are called

changes in entropy   written ΔS

Change in entropy is just the scaled-down change in the logarithm of the number of molecular arrangements of all kinds.

A 1 g of carbon dioxide (CO_2)   D 1 g of methane (CH_4)
B 1 g of hydrogen (H_2)   E 1 g of helium (He)
C 1 g of chlorine (Cl_2)

3 The gas phosphine contains phosphorus and hydrogen only. At 25 °C and 1 atm, 34 g of phosphine have a volume of 24.30 cubic decimetres; this volume yields 36.00 cubic decimetres (at 25 °C and 1 atm) of hydrogen on decomposition. Deduce the formula of phosphine. Explain how you obtained your answer.

4 Devise an alternative form of the Ideal Gas Law, using the symbols: p, V, R, T, m, M; where m = the mass of gas and M = the relative molecular mass of the gas.

5 Using the Ideal Gas Equation $pV = nLkT$, work out the pressure required, in newtons per square metre, to keep one mole of gas at 300 K in a volume of 0.01 cubic metre. ($L = 6.02 \times 10^{23}\,\text{mol}^{-1}$, $k = 1.38 \times 10^{-23}\,\text{J K}^{-1}$.)

6 A volatile liquid compound was found to be composed of carbon and hydrogen in the ratio of 1 mole of carbon atoms to 2 moles of hydrogen atoms. 0.124 g of the liquid on evaporation at 100 °C and 1 atm gave rise to 45 cm³ of vapour.
a What is the relative molecular mass of the liquid?
b What is the molecular formula of the liquid?

7 The following experiment is to determine the relative molecular mass of an acidic gas X.

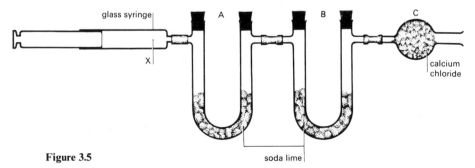

Figure 3.5   soda lime

Some of X was drawn into a glass syringe and its volume noted. The syringe was then connected to two U-tubes, A and B, as shown in figure 3.5, and these to the tube C. The gas X was slowly forced out of the syringe. The combined mass of A and B was determined before and after the experiment.

Volume of X at $16\,^{\circ}$C and 1 atm $= 100\,\text{cm}^3$
Combined mass of A and B before experiment $= 72.640\,\text{g}$
Combined mass of A and B after experiment $= 72.912\,\text{g}$

a Calculate the relative molecular mass of X (you may assume that X behaves like an ideal gas, that is, 1 mole of X molecules occupies a volume of 22.4 cubic decimetres at s.t.p.).

b Suggest a reason why *two* soda lime tubes were used.

c Suggest a modification to the procedure which would find out whether the tube B was necessary.

d What is the purpose of C?

e What modification would you adopt to determine the relative molecular mass of ammonia?

8 $10\,\text{cm}^3$ of hydrogen fluoride gas react with $5\,\text{cm}^3$ of dinitrogen difluoride gas (N_2F_2) to form $10\,\text{cm}^3$ of a single gas. Which of the following is the most likely equation for the reaction? Show how you reach your decision.

A $HF + N_2F_2 \longrightarrow N_2HF_3$
B $2HF + N_2F_2 \longrightarrow 2NHF_2$
C $2HF + N_2F_2 \longrightarrow N_2H_2F_4$
D $HF + 2N_2F_2 \longrightarrow N_4HF_5$
E $2HF + 2N_2F_2 \longrightarrow 2N_2HF_3$

9 1 volume of the gaseous element X combined with 1 volume of the gaseous element Y to form 2 volumes of a gaseous compound Z. Z is the only product. Which of the following statements conflicts with this evidence? Show how you reach your decision.

A When 1 molecule of X reacts with 1 molecule of Y, an even number of molecules of Z is formed.

B 1 molecule of Z could contain an even number of atoms.

C 1 mole of X reacts with 1 mole of Y.

D Both 1 molecule of X and 1 molecule of Y could contain an odd number of atoms.

E Both 1 molecule of X and 1 molecule of Y could contain an even number of atoms.

10 Figure 3.6 shows an apparatus which may be used for determining the relative molecular masses of gases. It consists of a gas syringe held vertically with a glass tube attached to it. The glass tube has a piece of aluminium foil glued across its end. The foil has a small hole in it,

Atomic structure

There is no simple model of atoms which enables us to understand all their observed properties. This is partly because the real nature of the atom is complex, and partly because our knowledge is still incomplete in spite of modern discoveries. But in order to discuss the behaviour of atoms it is necessary to suggest a model by which they may be pictured. This Topic describes such a model, and indicates the supporting evidence.

For convenience, we may consider an atom in two principal parts: the nucleus and the electrons.

4.1
THE NUCLEUS
What is the evidence for atomic nuclei?

In 1897 J. J. Thomson discovered the electron, and in 1899 he put forward a model of the atom consisting of rings of negatively charged electrons embedded in a sphere of positive charge so that a neutral atom resulted. As the mass of the atom was considered to be due only to the electrons, there had to be 1800 electrons in the hydrogen atom.

A few years later Dr H. Geiger, working in Manchester under the guidance of Professor Rutherford, discovered that when α-particles are fired at a thin metal foil in an evacuated container, most of the particles which penetrate the foil do so in a course which is undeviated or only very slightly deviated. In this experiment, a narrow pencil of α-particles from a source fell on a zinc sulphide screen, and the distribution of the scintillations on the screen was observed when different metal foils were placed in the path of the particles. The general arrangement of the apparatus is shown in figure 4.1.

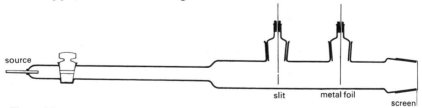

Figure 4.1
The main components of Geiger's apparatus.

On the Thomson model, it was to be expected that most of the α-particles should penetrate in a course that was undeviated or little deviated. Professor Rutherford takes up the story from here.

' One day Geiger came to me and said, "Don't you think that young Marsden, whom I am training in radioactive methods, ought to begin a small research?" Now I had thought that, too, so I said, "Why not let him sée if any α-particles can be scattered through a large angle?" I may tell you in confidence that I did not believe that they would be, since we knew that the α-particle was a very fast, massive particle, with a great deal of energy, and you could show that if the scattering was due to the accumulated effect of a number of small scatterings the chance of an α-particle's being scattered backwards was very small. Then I remember two or three days later Geiger coming to me in great excitement and saying, "We have been able to get some of the α-particles coming backwards...." It was quite the most incredible event that has ever happened to me in my life. It was almost as incredible as if you fired a 15-inch shell at a piece of tissue paper and it came back and hit you. On consideration, I realized that this scattering backwards must be the result of a single collision, and when I made calculations I saw that it was impossible to get anything of that order of magnitude unless you took a system in which the greater part of the mass of the atom was concentrated in a minute nucleus. It was then that I had the idea of an atom with a minute massive centre carrying a charge. I worked out mathematically what laws the scattering should obey, and I found that the number of particles scattered through a given angle should be proportional to the thickness of the scattering foil, the square of the nuclear charge, and inversely proportional to the fourth power of the velocity. These deductions were later verified by Geiger and Marsden in a series of beautiful experiments....'

This extract is taken from the 1936 essay by Ernest Rutherford, 'The development of the theory of atomic structure', in *Background to modern science*, J. Needham and W. Pagel (eds), The MacMillan Company, New York, 1938.

The apparatus used by Geiger and Marsden, shown in figure 4.2, was similar to that used by Geiger in the earlier experiments except that deviations of the α-particles through large angles could be observed by means of the movable zinc sulphide screen.

Figure 4.3 illustrates diagrammatically Rutherford's interpretation of this experimental result.

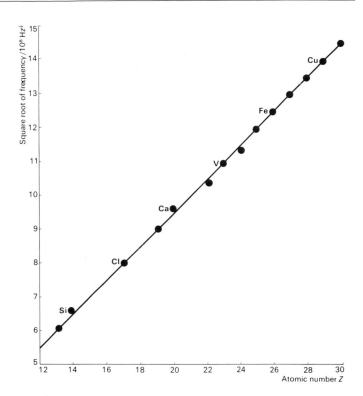

Figure 4.4

What particles are present in the nucleus?

In 1913 J. J. Thomson discovered that a given element can possess atoms with different masses, and the principle of his experiment was developed from 1919 onwards by Aston and others for use in the accurate comparison of these masses. This apparatus, the mass spectrometer, will be described later. As the atoms of a given element must all have the same atomic number (otherwise they would be atoms of a different element), they must have the same number of electrons and the same charge on the nucleus. To account for the different masses, Rutherford suggested that the nuclei contained the same number of positively-charged particles, protons; and he further suggested the existence of a particle with no charge but with the same mass as the proton which he named the neutron. Neutrons were first observed experimentally in 1932 by Chadwick.

Three different types of hydrogen atom are now known, with masses in the ratio 1:2:3. The nuclei are represented in the next diagram.

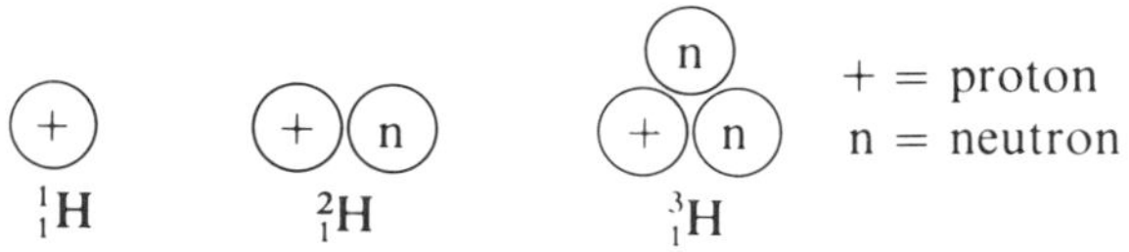

Each of these atoms contains one proton, and is therefore hydrogen. Each differs from the others by having a different number of neutrons. Atoms of an element which differ only in the number of neutrons that they contain are known as *isotopes*.

A convention is adopted by which the number of protons and neutrons in atoms may be readily shown on paper. The convention is used in the diagram of the hydrogen isotopes. The number of protons (that is, the atomic number) is placed at the bottom left hand of the symbol for the element, and the total number of protons and neutrons (that is, the mass number) is placed at the top left hand of the symbol. The number of neutrons is obtained by subtraction.

Two of the isotopes of lithium are shown below:

Each of these atoms would have three electrons outside the nucleus. Why have they not been represented on this diagram, assuming that the same scale has been maintained? In addition to the reason which you should have suggested, it is very difficult to say precisely where an electron is at any given instant.

How dense is the nucleus of the atom?

If the nucleus is so small and yet contains most of the mass of an atom, it must be very dense.

Calculate the density of the nucleus of a fluorine atom in $g\,cm^{-3}$ and in tonnes cm^{-3} from the following data.

The nucleus of a fluorine atom has a radius of approximately 5×10^{-13} cm and a mass of approximately 3.15×10^{-23} g. (1 tonne or metric ton = 1 Mg = 1000 kg.)

What does the result tell us about the strength of the forces holding the nucleus together?

Do you think that these forces penetrate very far outside the nucleus?

4.2
THE ELECTRONS

It is the electrons of an atom which are involved in chemical changes, not the protons and neutrons; this is to be expected because the electrons constitute the outer part of an atom. When one atom combines with another, either one or more electrons are transferred from one atom to the other, or electrons may be shared between the two atoms. The final arrangement need not concern us at the moment; it will, however, be clear that it is important to know the amount of energy which is needed to remove an electron from the atom if we are to understand the energy changes involved in chemical bonding. This energy is known as the *ionization energy*.

The first ionization energy of an element is the energy required to remove one mole of electrons from one mole of atoms of the element in the gaseous state, to form ions.

$$M(g) \longrightarrow M^+(g) + e^-; \qquad \Delta H = \text{ionization energy}$$

The ionization energy can be measured in electronvolts or kilojoules per mole; in this book we shall use the second of these units ($kJ\,mol^{-1}$). The reasons for using this unit are discussed in Topic 6.

Several different methods are available for determining the ionization energy of an element, and two will be considered here. One is for elements in the gaseous form and involves bombardment of the gas atoms by electrons; the other can be used for elements in either the gaseous or solid form and involves a study of their spectra.

Emission spectra of elements

When atoms of an element are supplied with sufficient energy they will emit light. This energy may be provided in several ways. If the element is a gas it may be placed in an electric discharge tube at low pressure; neon signs work on this principle. Certain easily vaporized metals also emit light under these conditions; examples are the bluish-white street lamps which are mercury discharge tubes, and the yellow street lamps which are sodium discharge tubes.

The energy may also be supplied by a flame; many metals and their salts when vaporized in a flame emit light. A further method which is particularly appropriate for metals and alloys is to make the element part of one electrode in an electric arc discharge.

When the light which is emitted is examined through a spectroscope, it is found not to consist of a continuous range of colours like part of a rainbow, but to be made up of discrete lines of colour. This type of spectrum is known as a *line emission spectrum*.

If you have done experiment 2.7, on page 37, you will have seen examples of these line emission spectra. Each element has its own characteristic set of lines, and these enable elements to be identified by examination of their spectra. Indeed, spectroscopic examination of the Sun revealed the existence there of an element which at that time had not been discovered on Earth; it was named helium, from the Greek word *helios*, meaning the Sun.

The photograph in figure 4.5 shows the emission spectrum of atomic hydrogen, in the visible region. (A coloured photograph of the hydrogen spectrum is given in the *Book of data*.) Each line corresponds to a given frequency, and the lines fit into a series, known as the *Balmer series*.

Frequency/10^{14} Hz

Figure 4.5
Labelled photograph of spectrum of atomic hydrogen, visible region.
Photograph, Dr W. F. Sherman, Department of Physics, King's College, London.

Balmer series of lines in the spectrum of atomic hydrogen

Frequency, $v/10^{14}$ Hz
$(1\,\text{Hz} = 1\,\text{s}^{-1})$

Red   α 4.568
　　　 β 6.167
Violet γ 6.907
　　　 7.309
　　　 7.551
　　　 7.709
　　　 7.817
　　　 7.894

What do you notice about the spacing of the lines?

Plot a graph of v against $1/n^2$ where
$n = (2 + 1)$ for the first line in the visible region of the spectrum (α)
　　 $(2 + 2)$ for the second line (β)
　　 $(2 + 3)$ for the third line (γ)
　　 etc.

What do you think has happened to the lines, and also to the hydrogen atom when $1/n^2 = 0$?

What is the interpretation of the two main features of the spectrum of hydrogen, first the discrete lines, and second the fact that the lines come closer together until they coalesce?

In order to explain these observations it is necessary to assume that an electron in an atom can exist only in certain energy levels; it cannot possess energy of intermediate magnitude. Energy is required to promote an electron from a low energy level to a high one (excitation); and if an electron falls from a high level to a lower one energy is released. This energy is not released over a continuous wide band of frequencies, but at a unique frequency, and we say that a quantum of radiation (light) has been emitted. For each electron transition, therefore, an associated quantum of radiation is emitted.

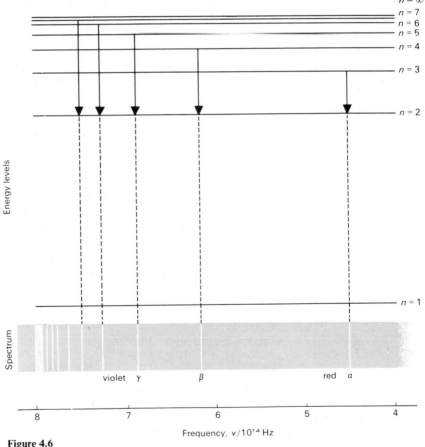

Figure 4.6
Energy level diagram, showing the origin of the lines in the visible portion of the hydrogen spectrum (not to scale).

Suppose in figure 4.6 that energy level $n = 3$ is associated with an energy E_3, and the level $n = 2$ with an energy E_2; then if an electron is transferred from n_3 to n_2 an amount of energy

$$E_3 - E_2 = \Delta E$$

is released. It has also been found necessary to assume that there is a relationship between a quantum of energy, ΔE, and its frequency, v, if it appears as radiation, such that

$$\Delta E = \text{constant} \times v$$

This relationship has been determined experimentally and the constant is called the Planck constant, h. If ΔE is measured in joules, and v in seconds^{-1}, h has the value 6.6×10^{-34} J s.

Since ΔE for the change from E_3 to E_2 is always the same in a given atom, and h is a constant, v must also be a constant; that is, the radiation always has the same energy and is always of the same frequency for this particular electron transition. Thus a discrete line appears in the spectrum.

Since the spectral lines converge and finally come together it is assumed that the electronic energy levels in an atom also converge and finally come together. This is shown in figure 4.6.

Transitions of an electron from various energy levels in hydrogen to energy level $n = 2$ involve energy changes such that the radiation emitted appears in the visible part of the spectrum. As mentioned earlier, this series of lines is known as the Balmer series. If, however, the transitions are to the $n = 1$ level, more energy is released, and the lines appear in the higher energy range of the spectrum, that is, in the ultra-violet region, and form the *Lyman series*. Similarly, if transitions occur from high levels to the $n = 3$ level, much less energy is released and the lines appear in the low energy region of the spectrum, that is, in the infra-red region. These lines form the *Paschen series*.

Make a copy of figure 4.6 in your notebook, and draw in transitions from high levels to either the $n = 1$ level or the $n = 3$ level. Then draw in schematically the general position of the spectral lines that would be produced.

If sufficient energy is supplied to an atom to promote an electron from one energy level to the highest possible one and just beyond it, then the electron is able to escape, and the atom becomes an ion. It should therefore be possible to determine the ionization energy of an element from its spectrum. This can be done if the frequency can be determined at which the converging spectral lines actually come together. This frequency is known as the 'convergence limit'.

The table below gives the frequency of lines in the ultra-violet spectrum of atomic hydrogen, which form the Lyman series.

Lyman series of lines in the spectrum of atomic hydrogen

Frequency, $v/10^{15}$ Hz

2.466
2.923
3.083
3.157
3.197
3.221
3.237
3.248

You will notice that the values of the frequencies come closer together, and once again converge to a limit. Since the ultra-violet spectrum represents electrons making transitions to the lowest energy level, $n = 1$, the convergence limit represents the energy required to ionize a hydrogen atom with its electron in the lowest level; hence it may be used to find the ionization energy of hydrogen. Find the frequency of the convergence limit as follows. Work out the difference in frequency, Δv, between successive lines, and plot a graph of Δv against the frequency v. Then extrapolate the curve to the point where the difference in frequency, Δv, becomes equal to 0 and read off the frequency. It does not matter whether you use the value of the higher or the lower frequencies for plotting v, as long as you are consistent in your choice. If two curves are plotted, one using the higher values of the frequencies and the other using the lower values, both on the same graph, it will be easier to estimate the value of the frequency when $\Delta v = 0$.

Convert this frequency into $kJ\,mol^{-1}$ using the Planck constant given above. Compare the value for the ionization energy of hydrogen that you obtain with that given in the *Book of data*.

Note: The units of the Planck constant are joule second. Multiplying the frequency that you obtain from your graph by this constant (6.6×10^{-34} J s) gives the value in *joules* of the ionization energy of a *single atom* of hydrogen. To convert this figure into the ionization energy of hydrogen in kilojoules per mole, the units given in the *Book of data*, first multiply it by the Avogadro constant L ($6.06 \times 10^{23}\,mol^{-1}$) to give joules per mole, and then divide it by 1000, to give kilojoules per mole.

Electron bombardment of gases

Another method by which the ionization energies of certain elements can be found is by the bombardment of atoms of gaseous elements by electrons.

A closed tube containing two electrodes and the gas under low pressure is required, together with some means of heating the cathode within the tube to a dull red heat. A potential difference is applied across the electrodes, the voltage is varied, and the resulting current is measured. The arrangement is shown in figure 4.7.

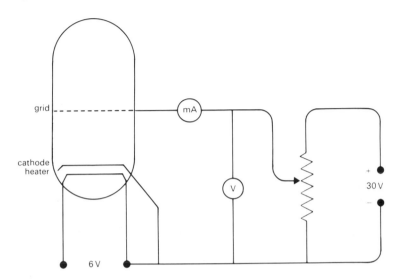

Figure 4.7
Circuit for determining the ionization energy of a noble gas.

Voltages in the range 0–30 V are required, and the current may be in the region of microamperes, or possibly milliamperes.

As the voltage is steadily increased, the current increases in proportion, showing that the almost empty space in the tube between the electrodes, across which the electrons are travelling, obeys Ohm's Law. However, at a certain voltage, which is different for each gas contained in the tube, the current begins to increase at a much more rapid rate. A graph of voltage against current is of the type shown in figure 4.8, which was obtained using a tube containing helium at low pressure.

The explanation of this behaviour is as follows. When the cathode is heated electrons are emitted, and are attracted to the anode by the potential difference applied across these electrodes. As they travel from one electrode to the other, they may strike a gas atom, and if they do, they are merely deflected in their path, and may be slowed down.

As the potential difference is increased, however, these electrons gain more and more energy until, at a certain voltage, their energy is so great that when they collide with a gas atom, they are able to knock an electron out of that atom. That is, they are able to *ionize* the atom. There are then two free electrons

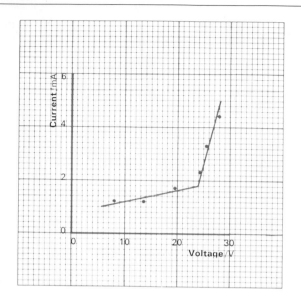

Figure 4.8

where previously there was only one, and so a greater current is recorded than would otherwise be the case.

A commercial source of such tubes exists. Older radio and television sets include *valves* in their circuitry, and some of these valves can be used to measure the first ionization energy of the gases with which they are filled. The heaters in these valves usually require 6 volts, and either form the cathode or are electrically connected to it. There may be a single anode (in a 'diode') or several other electrodes (as in 'triodes' etc.). In this experiment, the 'grid' can be used as the anode.

You may be able to do an experiment to measure the ionization energy of one of the noble gases by such a method, depending upon the availability of suitable radio valves.

The first ionization energy of an element such as argon can be found from the potential difference in volts at which ionization of the gas takes place, by multiplying this potential difference by 96.3. This figure arises as follows.

As 1 joule of energy has to be expended when 1 coulomb of charge passes through a potential difference of 1 volt,

$$1\,V = 1\,J\,C^{-1}.$$

When one electron (charge $= 1.6 \times 10^{-19}$ C) is accelerated through a potential difference of 1 V, it acquires $1.6 \times 10^{-19} \times 1$ joule of energy. When 1 mole of electrons is accelerated through a potential difference of 1 V, it acquires

$$1.6 \times 10^{-19} \times 6.02 \times 10^{23} \, J$$
$$= 9.63 \times 10^4 \, J$$
$$= 96.3 \, kJ$$

The amount of energy possessed by electrons when accelerated by a potential difference of 1 volt is therefore $96.3 \, kJ \, mol^{-1}$. So the energy that 1 mole of electrons had when it ionized 1 mole of argon atoms can be calculated from the accelerating voltage by multiplying the voltage by 96.3.

What can ionization energies tell us about the arrangement of electrons in atoms?

The discussion of the ionization of an atom has so far considered the removal of one electron only; but if an atom containing several electrons is treated with sufficient vigour, then more than one electron may be removed from it. A succession of ionization energies is therefore possible. These may be determined, principally from spectroscopic measurements; a table of successive ionization energies for a number of elements is given in the *Book of data*.

1 Using this table, attempt to plot for sodium a graph of number of electrons removed against the appropriate ionization energy. What do you notice, first about the general trend in values, and second about their magnitude? Why does the general trend (increase or decrease in values) occur?
Now plot a graph of the logarithm of the ionization energy, on the vertical axis, against number of electrons removed, on the horizontal axis.
Does this give any information about groups of electrons which can be removed more readily than others? How many electrons are there in each group?

2 Using the same table, study the change in the first ionization energy of the elements. For the first twenty elements plot the value of their first ionization energy, on the vertical axis, against their atomic number, on the horizontal axis. When you have plotted the points, draw lines between them to show the pattern, and label each point with the symbol for the element.
Where do the alkali metals lithium, sodium, and potassium appear?
Where do the noble gases occur?
Do you notice any groups of points in the pattern? How many elements are there in each group? Do the numbers bear any relation to the numbers of electrons in any pattern you may have found for the successive ionization energies for sodium?

3 What interpretations can be placed on these results?
In the successive ionizations of sodium, one electron needs much less energy

for its removal than did the others, and it must therefore have been in a high energy level; eight electrons required much more energy, and must have been in a lower energy level; two electrons must have been in a still lower energy level. The lowest energy level is called the $n = 1$ level.

Thus, for sodium there would appear to be:

2 electrons in the $n = 1$ energy level
8 electrons in the $n = 2$ energy level and
1 electron in the $n = 3$ energy level.

We can represent this on an energy level diagram, as in figure 4.9.

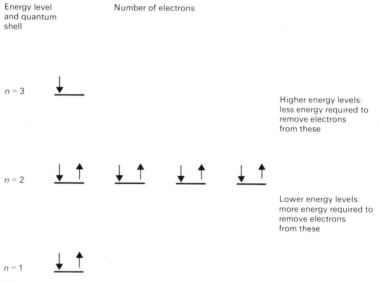

Figure 4.9
Energy levels of electrons in a sodium atom.

The two electrons in the $n = 1$ energy level are situated *most* of the time closer to the nucleus than the other electrons, and they are said to be in the first *quantum shell*. The eight electrons in the $n = 2$ energy level spend much of their time further from the nucleus, and are said to be in the second quantum shell. The single electron in sodium spends much of its time further still from the nucleus and is said to be in the third quantum shell.

Thus there are two ways of looking at electrons in atoms: from the point of view of their energy level, $n = 1, 2, 3, 4$, etc., and from the point of view of how far from the nucleus they are on average, that is, in the first second, third, or fourth, etc., quantum shell.

The electrons in figure 4.9 have been represented by arrows. When an energy level is half full the next electrons pair up with existing ones. Electrons behave as though they had the property of spin, and paired electrons must have their spins in opposite directions; this is represented by up and down arrows. The reasons why we believe that electrons behave as though they were spinning, and that the spins of paired electrons are opposed, are complicated, and it is not necessary to discuss them here. The evidence for this comes from a more detailed examination of line spectra.

If you had plotted the successive ionization energies for potassium, the pattern would have been 1 electron most easily removed, followed by 8 more difficult, followed by 8 even more difficult, followed by 2 extremely difficult to remove. How many electrons do the $n = 1$, $n = 2$, $n = 3$, and $n = 4$ energy levels and quantum shells hold in potassium?

Does a similar pattern show in the graph of first ionization energies which you plotted for 20 elements?

You will notice from this latter graph that the groups of eight are made up of groups of (2, 3, and 3) points on the curve. This indicates that the eight electrons are not all exactly the same as far as their energies are concerned. From this type of evidence, and also from studies of spectral lines, it has been concluded that the energy levels are split so that the $n = 2$ level has two electrons in a sub-level known as 2s (slightly more difficult to remove) and six electrons in a sub-level known as 2p (slightly less difficult to remove). This is shown in figure 4.10.

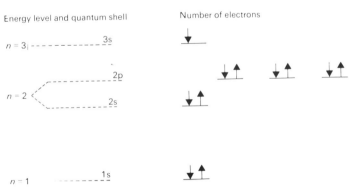

Figure 4.10
Energy levels of electrons in a sodium atom, showing sub-levels.

From similar evidence, it has been concluded that the $n = 3$ level is split into s, p, and d, and the $n = 4$ and $n = 5$ levels into s, p, d, and f sub-levels. All the s sub-levels can contain up to two electrons, the p sub-levels six, the d sub-levels ten, and the f sub-levels fourteen. The arrangement of the $n = 1$ to $n = 4$ energy levels is shown in figure 4.11.

4.3
ELECTRONIC STRUCTURE AND TRENDS ACROSS A PERIOD

Draw up a table in your notebook with the headings shown below:

Element	Li	Be	B	C	N	O	F	Ne
Atomic radius /nm								
First ionization energy/kJ mol^{-1}								
Element	Na	Mg	Al	Si	P	S	Cl	Ar
Atomic radius /nm								
First ionization energy/kJ mol^{-1}								

Fill in the values of the atomic radius, and the first ionization energy, of each element of these two periods of the Periodic Table. For the atomic radius of each of the elements except the noble gases, you should use values of the *covalent radius*, r_{cov}, taken from the table of atomic radii in the *Book of data*. For the noble gases you should use the only values quoted, which are of the *van der Waals radius*, r_v. The first ionization energies are given in the *Book of data*.

When you have completed your table, refer to it as you read the next two paragraphs and answer the questions. Write your answers in your notebook after the table, in such a way as to make clear what you are answering.

1 Atomic radius

As can be seen from the table of atomic radii in the *Book of data*, a number of different atomic radii can be distinguished. When making comparisons between elements, take care to select the appropriate values.

The *covalent radius* of an atom in an element is half the distance between the centres of two adjacent atoms in a close-packed structure or in a molecule.

The *van der Waals radius* is half the distance between the centres of two atoms in adjacent molecules.

The difference between these two quantities is discussed further in Topic 10, and is illustrated in figure 10.18 (page 363). Atoms of the noble gases do not form molecules, and so they can only have van der Waals radii.

How does atomic radius change across a period of the Periodic Table? Why is this?

2 Trends in ionization energies

As a broad trend how does the first ionization energy change across a period of the Periodic Table? Why is this?

How does the first ionization energy change on going down a group in the Periodic Table? Why is this?

3 Chemical similarities

The chemical similarities existing among members of a group of elements arise because of the similar configurations of the outer electron shells of their atoms.

The lack of reactivity of the noble gases is largely due to their very high ionization energies.

4.4
THE ACCURATE DETERMINATION OF RELATIVE ATOMIC AND MOLECULAR MASSES

The most accurate method of determing relative atomic and molecular masses is by use of the mass spectrometer. Figure 4.14 shows how it functions.

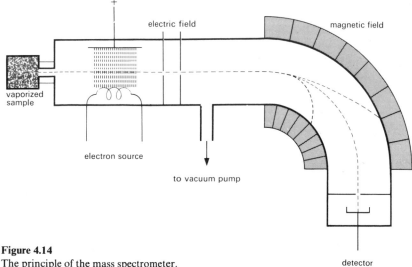

Figure 4.14
The principle of the mass spectrometer.

Five main operations are performed by the spectrometer:
1 The sample is vaporized.
2 Positive ions are produced from the vapour.
3 The positive ions are accelerated by a known electric field.
4 The ions are then deflected by a known magnetic field.
5 The ions are then detected.

First let us consider the determination of the relative atomic mass of an element. A stream of the vaporized element enters the main apparatus, which is maintained under high vacuum. The atoms of the element are bombarded by a stream of high-energy electrons which, on collision with the atoms, knock electrons out of them and produce positive ions. In most cases single electrons will be removed from atoms of the element

$$E(g) \longrightarrow E^+(g) + e^-$$

although in some cases more electrons may be removed, for example,

$$E^+(g) \longrightarrow E^{2+}(g) + e^-$$

The positive ion stream passes through holes in two parallel plates to which a known electric field is applied, and the ions are accelerated by this field. They then enter a region to which a magnetic field is applied, and they are deflected by it.

For given electric and magnetic fields only ions with the same charge and one particular mass will reach the detector at the end of the apparatus, all other ions having hit the walls of the instrument. By gradually increasing the strength of the magnetic field, ions of increasing masses may be brought successively to the detector. Their masses are calculated from the known applied fields, and their relative abundance is found from the relative magnitudes of the current produced in the detector.

Figure 4.15 shows a mass spectrometer trace for naturally occurring lead, as an example of the type of trace which is obtained. It will be seen that the mass spectrometer gives the relative abundance of the various isotopes of the element.

Note that the horizontal axis is labelled 'Mass/charge ratio'. This is because ions of the same mass but with different charges give separate traces in the mass spectrum. For ions carrying a single charge the mass/charge ratio is equal to the isotopic mass.

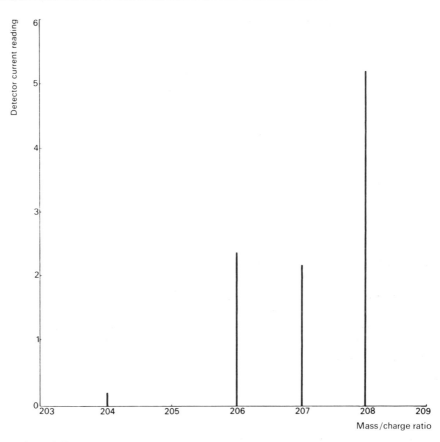

Figure 4.15
Trace obtained from the mass spectrometer, for naturally occurring lead.

From figure 4.15 the relative abundances can be seen to be:

Isotopic mass	Relative abundance	% relative abundance
204.0	0.2	2
206.0	2.4	24
207.0	2.2	22
208.0	5.2	52
	10.0	100

From these values the atomic mass of naturally occurring lead can be worked out as follows.

In 100 atoms of naturally occurring lead there will be, on average, 2 atoms of isotopic mass 204.0, 24 of 206.0, 22 of 207.0, and 52 of 208.0. If we find the total mass of all of these 100 atoms we may find the average mass by dividing by 100.

Isotopic mass	Number of atoms in 100 atoms of mixture	Mass of isotopes in 100 atoms of mixture
204.0	2	408
206.0	24	4 944
207.0	22	4 554
208.0	52	10 816
		20 722

$$\text{Average mass of 1 atom} = \frac{20\,722}{100} = 207.2 \text{ atomic mass units}$$

This is the *relative atomic mass*, the reference standard for which is the mass of one atom of the ^{12}C isotope, which is taken as exactly 12 units of atomic mass (see Topic 1, section 1).

Use the table of 'Atomic masses, sizes, and abundances' in the *Book of data* to work out the relative atomic masses of naturally occurring chlorine and of magnesium. Record the working, and the result, in your notebook.

Examine the abundance of the isotopes of $_{52}Te$ and the relative atomic mass of tellurium. Compare this with the relative atomic mass of $_{53}I$ and comment on their positions in the Periodic Table and their relative atomic masses. Do the same for $_{18}Ar$ and $_{19}K$.

We can now consider the determination of relative molecular masses.

The use of the mass spectrometer for the accurate determination of relative atomic masses can be applied to the problem of determining relative molecular masses accurately, particularly for volatile compounds such as most carbon compounds.

Relative molecular masses can be found by the determination of the vapour densities of the compounds, as explained in Topic 3. However, the relative molecular masses of volatile compounds are now determined in industrial and other research laboratories by using a mass spectrometer.

For the determination of relative molecular mass, the compound under investigation is injected into the instrument as a vapour. (It must, of course, be stable at whatever temperature is needed to turn it to a vapour at about $10^{-4}\,N\,m^{-2}$, the pressure inside the instrument.) High velocity electrons then bombard the molecules and produce a variety of positively-charged ions.

In the case of dodecane, $C_{12}H_{26}$, if one electron is knocked out of the molecule by the bombardment, the $C_{12}H_{26}^{+}$ ion will be formed, and the detector will show the presence of an ion of mass number 170. The electron bombardment, however, not only has the effect of knocking out electrons from the molecules; it may also break the molecules into smaller fragments, such as the $C_6H_{13}^+$, $C_5H_{11}^+$, $C_4H_9^+$ and other ions (see figure 4.16). Again, the horizontal axis is labelled mass/charge ratio. More than one electron may be removed from some fragments, and this will give a different mass/charge ratio and thus a separate trace in the spectrum.

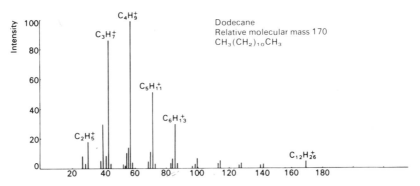

Figure 4.16   Mass/charge ratio
The mass spectrum of dodecane.

The ion detected to have the highest mass, the 'parent ion', normally indicates the relative molecular mass of the compound. From this, and from a knowledge of the elements present, some idea of the molecular formula can be obtained by reference to tables of masses which have been compiled for the purpose. For example, if the mass of the parent ion was 200, and the compound contained C, H, and O only, possible molecular formulae would include $C_{10}H_{16}O_4$, $C_{11}H_4O_4$, $C_{11}H_{20}O_3$, and six others. It is usually possible to find which species is actually present by using a high-resolution instrument giving a higher degree of accuracy, for there are small variations in relative molecular mass between each of the possible examples. For example, using the values of the most abundant isotopes,

$$^{16}O = 15.99491$$
$$^{12}C = 12.00000$$
$$^{1}H = 1.007829$$

the relative molecular masses of the formulae given above are

$$C_{10}H_{16}O_4 = 200.1049$$
$$C_{11}H_4O_4 = 200.0110$$
$$C_{11}H_{20}O_3 = 200.1413$$

for molecules made up of atoms of the stated isotopes. Thus, if the high resolution mass spectrum showed that the parent ion had an exact isotopic mass of 200.011, one can be sure that the molecular formula of the compound is $C_{11}H_4O_4$. The presence of other isotopes will, of course, lead to small numbers of molecules having slightly different masses, and hence to low intensity peaks in the mass spectrum.

Having found the molecular formula, some idea of the structure of the compound can be obtained from the ions of smaller mass, caused by the break-up of some of the original molecules under high velocity electron bombardment.

Alternatively, as the nature and proportions of these different fragments are characteristic of the original compound, the spectrum obtained can be used as a sort of 'fingerprint' for identification purposes, by comparison with the spectrum obtained from a sample of the authentic compound.

Once the molecular formula has been found, the molecular structure can be determined. This can be done by examination of, for example, the infra-red spectrum or nuclear magnetic resonance spectrum, as will be explained in Topic 7.

4.5
THE CHANCE BEHAVIOUR OF ENERGY

In Topic 3 we began to see how thinking about numbers of arrangements of moving molecules in a gas could help us to think about chemical and physical change.

Here we shall see how these ideas can be extended to counting *states of energy* of particles.

In this Topic we have seen that the energy of an atom is *quantized*. It comes in lumps called quanta. This is true for the energies of electrons in an atom. It is also true for vibrations of an atom in a solid, and for other movements of atoms.

A good way to see that vibrational energy is quantized is to look at the spectrum of white light through purple iodine vapour, as described in the next experiment.

EXPERIMENT 4.5
An illustration of the quantization of energy

Carefully heat a few small crystals of iodine placed at the bottom of a $150 \times 16\,mm$ test-tube with a low, colourless Bunsen burner flame, so as to vaporize the iodine without driving it out of the open end of the test-tube.

WARNING: Wear safety glasses and do not overheat the test-tube. Iodine vapour is dangerous; be careful that it does not get into your eyes, where it may crystallize painfully.

Hold the test-tube with the iodine vapour in front of a low-voltage lamp. Hold a diffraction grating close to your eye, and examine the spectrum of the light which passes through the iodine. Iodine, I_2, has a diatomic molecule which can vibrate. It can absorb light energy and increase its energy of vibration. Where this happens, the absorbed light leaves a dark band in the spectrum. If you look at the spectrum, you see that it consists of a series of bands arranged like the rungs of a ladder. These bands correspond to the steps in the quantized vibrations of I_2 molecules.

Now we want to see how to count numbers of ways of sharing quantized energy. We will take just one very simple case: an assembly of oscillating molecules making up a crystal. We take this example because experiment and theory both say that the energy levels are very simple in this case: just a ladder of equally spaced rungs (figure 4.17).

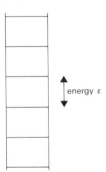

Figure 4.17

We shall write ε for the difference in energy between rungs, and so also for the size of energy unit (quantum) which can be exchanged.

Now suppose we have two such oscillating molecules and they both happen to have energy 2ε above the zero-point energy.

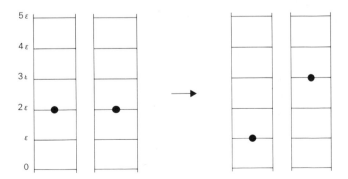

Figure 4.18

Let the energy be constant, but let them exchange energy between themselves. One thing that could happen is that the first molecule gives ε to the second, as shown in figure 4.18. It drops a rung, and the second molecule rises a rung.

What are *all* the ways the energy can be shared out?

Molecule 1	Molecule 2
0	4ε
1ε	3ε
2ε	2ε
3ε	1ε
4ε	0

There are just five different ways to share the energy. Now let them share one more unit of energy, total 5ε. Then we get

Molecule 1	Molecule 2
0	5ε
1ε	4ε
2ε	3ε
3ε	2ε
4ε	1ε
5ε	0

Now there are six ways. It is pretty obvious that giving some molecules more energy to share means more ways to share the energy. For two molecules, it is clear that the number of ways is one more than the number of shared units of energy (just look at the table). For more than two molecules, the counting is harder. For three molecules sharing three units of energy there are ten ways:

Molecule 1	Molecule 2	Molecule 3
3ε	0	0
0	3ε	0
0	0	3ε
1ε	1ε	1ε
1ε	2ε	0
2ε	1ε	0
1ε	0	2ε
2ε	0	1ε
0	1ε	2ε
0	2ε	1ε

For 10 molecules sharing ten units, there are 92 378 ways. For 100 molecules sharing 100 quanta, there are about 8×10^{59} possibilities. So, using the italic letter W to represent the number of possible ways as we did in Topic 3, we have

$W = 8 \times 10^{59}$

In Topic 3, we looked at changes in entropy, ΔS. Here we can be absolute, because we can count *all* the ways. We call $k \ln W$ the entropy of the molecules, that is,

$S = k \ln W$

The entropy is not very great because although 8×10^{59} is large, $\ln W$ is much less, 138 in fact, so that

$S = k \ln W = 1.38 \times 10^{-23} \ln(8 \times 10^{59}) = 1.9 \times 10^{-21} \, \text{J K}^{-1}$

For 10^{23} molecules sharing 10^{23} quanta the picture is very different. The logarithm of W is 1.4×10^{23} and the entropy is

$S = k \ln W = 1.9 \, \text{J K}^{-1}$

So for numbers of molecules like those in a few grams of matter (here $\frac{1}{6}$ mole), we can get entropies of the sort of size we had in Topic 3.

Energy doesn't care, either

You will remember that in Topic 3 we discussed the principle 'molecules don't care'. Now we can add another principle, namely, 'energy doesn't care, either'.

If energy can be shared by a set of molecules, it *will* be shared, and in all possible ways. This means that the energy that has warmed the molecules to (say) room temperature contributes to the entropy, because entropy involves the total number of ways of arranging both molecules and energy.

In fact, there is a direct parallel between diffusion (discussed in Topic 3) and a seemingly quite different question, namely things warming or cooling each other.

A hot cup of tea in a cool room soon cools. An ice cream in the same room warms up. Why? Just as before, energy doesn't care. The hot tea cools because passing energy to the room leads to more ways of sharing out the total. The ice cream gets energy from the room because taking in energy again leads to more ways of sharing.

So we can think of heat going from hot to cold as being a little like diffusion, with energy going from where it is relatively concentrated to where it is relatively less concentrated.

EXERCISE
A picture of shuffling quanta

With a new idea, it is often helpful to make a picture of what it means. This can be done for the shuffling of energy amongst oscillators quite easily, for cases where there are enough oscillators and enough energy for the system to be made more realistic.

 We will assume in this exercise that each oscillator possesses one quantum of energy. This makes things simple, and calculations show that it is a fairly accurate assumption at room temperature.

Hand version

Make a 6×6 grid of squares, each square representing one oscillator. Use counters to represent quanta of energy. A square with no counter on it is an oscillator at its lowest possible energy. Higher states are represented by having one, two, etc. counters on a square.

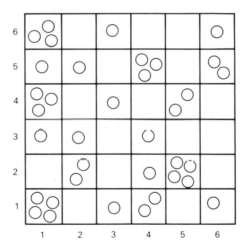

Figure 4.19
Typical appearance of a 6×6 grid after 100 moves.

 Start as you please. A simple starting point is to have 36 counters, one on each square. Throw two dice: let the numbers they show select a square. Pick up a counter from that square if it has one. If it does not have one, try again. Then throw the dice to select a square to receive the counter, and put it there. Then start again.

From time to time (after, say, every 20 complete 'pick-up' and 'put-down' moves), count the number of squares with 0, 1, 2, etc. counters. Plot histograms.

Distribution reaches an equilibrium, around which it varies quite a lot. The counters go on moving about, but the pattern of numbers of oscillators with different energy stays the same. Moreover, this pattern is *exponential* (figure 4.20).

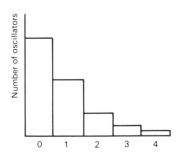

Figure 4.20

Another version is to have a 10 × 10 grid worked on by the whole class at the same time. Each member of the class is given a sheet of random numbers from 0 to 9 to use instead of dice. Or the school microcomputer can be programmed to generate random numbers over any desired range.

Computer version

With a computer doing the work, a much larger grid of oscillators can be used. 30 × 30 is a good choice. The essential thing is

    a to have a visual display of the grid and of 'quanta' moving about on it

    b to be able to get histograms from time to time showing the distribution.

Suitable programmes exist, and you may be able to use one to find the distribution after a large number of moves; 10000 is a good number to aim for.

Useful variations to play with a computer version are:

to vary the number of 'quanta' per oscillator

to vary the initial distribution of 'quanta', to see that the end result is the same

to exchange 'quanta' between two such systems, each with different numbers of 'quanta' per oscillator, to see thermal equilibrium established.

BACKGROUND READING
Atomic emission spectroscopy

Every element in the Periodic Table has a unique electronic structure. Excited atoms of a certain element will therefore give rise to a line spectrum which is different from that of any other element. Consequently, line spectra can be used to identify elements. Furthermore, the intensities of lines in an emission spectrum are proportional to the concentrations of the atoms present in a sample. Thus, a quantitative analysis is possible.

The instruments which were first developed for the purpose of carrying out this quantitative analysis were known as emission spectrographs. These recorded the spectra on a photographic plate and the intensities of the lines were subsequently measured, using a densitometer. The analyst needed to be skilled in interpreting spectra. The process, requiring the individual measurement of each line, was a lengthy one. The instruments now in use are known as automatic emission spectrometers. These employ photomultiplier tubes to measure the intensity of radiation at selected wavelengths. The photomultiplier detector converts radiant energy into electrical currents which, under controlled conditions, are proportional to the intensity of the radiation.

Atomic emission spectroscopy in its current form is not only capable of quantitative accuracy comparable with chemical methods but is also very much faster. It allows a large number of element concentrations to be measured at the same time. The automatic emission spectrometer is used in many different industries, for instance, in the analysis of pharmaceuticals, foods, agricultural products, oils, and metals, both ferrous and non-ferrous. The steel industry was the first to employ these instruments on a wide scale and it is probably still the biggest user.

The demand for steel in a technologically advanced society is enormous and extremely efficient production methods must be developed to meet it. This, in turn, means that there must be high capital investment in the plant and that it cannot be allowed to remain idle for longer than necessary during the process of steelmaking.

A modern steelmaking plant, with its oxygen converters and electric arc furnaces, consumes as much electricity as a medium-sized town. Before the molten steel can be poured, a sample of melt must be analysed to ensure that it is as specified. Keeping this analytical time down to a minimum can significantly lower the cost of production.

A skilled analyst, using 'wet chemical' methods, would require about 45 minutes to identify the major constituents in a sample of the melt. It would take longer still to determine trace elements. During this time, the cost of maintaining the electric furnace at the required temperature is not the only thing to be borne in mind. There is also the danger of atmospheric contamination of the material. Modern analytical instruments reduce this delay to a minimum.

Figure 4.21
Molten steel being poured.
Photograph, British Steel Corporation.

A good example of the kind currently in use is the Polyvac computer-controlled automatic emission spectrometer manufactured by Hilger Analytical. Typically, this will analyse a sample and provide a print-out of the percentage concentration of 20 or 30 elements (more if required) in less than a minute.

The Polyvac

A small sample of melt is taken from the furnace, cooled quickly, and immediately transferred to the laboratory. In some plants, this laboratory is situated as close

a video display unit or an automatic high-speed printer. The display unit, or the printer, can be placed close to the furnace, telling the man on the spot that the melt can be poured or that some adjustment must be made to its composition.

Excitation systems The way in which the sample is excited is crucial to the success of the analysis: the method is determined by the sample itself. Some of the most recently introduced excitation systems are the Hilger Jet Electrode, the Glow Discharge and the Inductively Coupled Plasma. The Jet Electrode makes possible the determination of certain steels and irons which were not previously amenable to spectrochemical analysis. The Glow Discharge has improved the analytical performance for other materials, while the Inductively Coupled Plasma system has brought liquid analysis within the scope of the Polyvac. The ICP, as it is called, is particularly suited to the analysis of oils. Oil companies, transport groups, and vehicle manufacturers use it for measuring the concentration of wear products in oils. In this way, it is possible to determine when an engine is in need of overhaul. This is particularly important from the point of view of safety, for transport groups, whether they are concerned with road, rail, or air, can establish exactly the right point at which an engine should be taken out of service.

The results The most time-consuming part of the analytical sequence is the drawing off of the sample, its preparation, and transport to the Polyvac spectrometer. This sequence will take several minutes, depending upon the efficiency with which it is carried out, whereas the actual analysis will be completed in less than a minute. Well organized, the whole operation, from the collection of the sample to the return of results to the furnaceman, can be completed in five minutes. This compares well with the forty-five minutes required by a skilled analyst. Furthermore, the latter would only be able to complete the analysis of the major constituents. The Polyvac will determine the trace elements as well. The number of elements to be analysed has very little effect on the analysis time.

While it is almost certain that a trained chemist will have overall responsibility for the laboratory where the Polyvac is, the instrument itself can be operated by relatively untrained people. Once the sample has been loaded, the computer sets the necessary parameters and then all the operator has to do is push a button.

SUMMARY

At the end of this Topic you should:

   1 be familiar with Geiger's and Marsden's experiment as evidence for the existence of atomic nuclei;

Analytical Results for Steels
Concentration % ± Standard Deviation

Al	.007	± .0002	Mn	.13	± .003	Si	0.29	± .0008
	.049	± .0015		.67	± .009		.29	± .004
				1.43	± .015		1.16	± .012
As	.010	± .0004						
	.032	± .001	Mo	0.21	± .0007	Sn	0.14	± .0005
				.26	± .003		0.34	± .0007
B	.0030	± .00009		3.07	± .04			
						Ti	.002	± .0002
C	.018	± .001	Nb	0.20	± .0008		.11	± .0015
	.54	± .006		.053	± .001			
	1.12	± .011				V	.006	± .0004
			Ni	.024	± .0006		.18	± .002
Co	.084	± .001		.38	± .004			
				1.52	± .015	W	.022	± .0015
Cr	.019	± .0007		8.12	± .050		6.24	± .07
	.36	± .004					18.03	± .11
	1.22	± .012	P	.0045	± .0003			
	20.67	± .09		.023	± .0005	Zr	.021	± .001
Cu	.016	± .0004	S	.010	± .0005			
	.30	± .004		.039	± .0012			

Figure 4.24
Print-out of results obtained with the Polyvac.
Photograph, Hilger Analytical Ltd.

    2 be familiar with Moseley's experiment relating the frequency of emitted X-rays with the atomic number of the element from which they are emitted;

    3 know that protons and neutrons are constituents of atomic nuclei;

    4 know the meaning of the term 'isotope';

    5 be aware that atomic nuclei have very high densities;

    6 know that the ionization energies of elements can be found

    a from a study of emission spectra

    b by the electron bombardment of gases
and be aware of the relationship $\varepsilon = h\nu$;

    7 know the arrangement of electrons in atoms, by shell and orbital, for the first 20 elements in atomic number order, and be able to write them, using notation of the type $1s^2 2s^2 2p^6$;

    8 be aware of the trends in atomic radii and ionization energy across a period of the Periodic Table;

    9 know the meaning of the terms 'covalent radii' and 'van der Waals radii' as applied to atoms;

    10 understand the principle of the mass spectrometer, know how it is applied to the determination of relative atomic and molecular masses, be able to work out relative atomic masses, given the proportion of the various isotopes, and be able to interpret mass spectra of simple compounds;

    11 have some awareness that entropy is concerned with the distribution of quanta of energy as well as the distribution of molecules;

    12 be aware of the operation and use of the atomic emission spectrometer.

i Which of these elements should have the largest atomic number? Give reasons for your answer.

ii In which group of the Periodic Table should the elements be placed? Give reasons for your answer.

Questions **6** to **9** refer to the following table of ionization energies $(kJ\ mol^{-1})$ of five elements (the letters are not the symbols for the elements).

Elements	1st ionization energy	2nd ionization energy	3rd ionization energy	4th ionization energy
A	520	7298	11815	–
B	578	1817	2745	11578
C	1086	2353	4621	6223
D	496	4563	6913	9544
E	590	1145	4912	6474

6 Which of the elements, when it reacts, is most likely to form a 3 + ion?

7 Which one of the following pairs of elements is likely to be in the same group of the Periodic Table?

B and E   D and E   C and E
A and D   B and C

8 Which of the elements would require the most energy to convert one mole of atoms into ions carrying one positive charge?

9 Which of the elements would require the most energy to convert one mole of atoms into ions carrying two positive charges?

10 Which of the following would require the *most* energy to convert them completely from the gaseous state into gaseous ions each carrying one positive charge?

A 1 mole of lithium atoms   D 1 mole of rubidium atoms
B 1 mole of sodium atoms   E 1 mole of caesium atoms
C 1 mole of potassium atoms

11 Which of the following would require the *least* energy to convert them completely from the gaseous state into gaseous ions each carrying one positive charge?

A 1 mole of lithium atoms   D 1 mole of carbon atoms
B 1 mole of beryllium atoms   E 1 mole of nitrogen atoms
C 1 mole of boron atoms

12 Natural silicon consists of a mixture of three isotopes and its
atomic number is 14.

Isotope	Isotopic mass	Percentage abundance by numbers of atoms
A	28.0	92.2
B	29.0	4.7
C	30.0	3.1

i In each of the isotopes how many neutrons are there in each atom?
How many protons are there?
ii Calculate the relative atomic mass of natural silicon. Show how you
arrive at your answer.
iii State, in the form of numbers and symbols, the energy levels of the
electrons in the isotope B.

13 In a distribution of quanta of energy among oscillators, using a rectangular
grid as a model in the way described in the text, the number of oscillators
n_0, n_1, n_2 etc. having no quanta, one quanta, two quanta, etc. were
$n_0 = 300$, $n_1 = 210$, $n_2 = 148$, and so on as shown in figure 4.26 on the next
page.

a What was the total number of oscillators in the system? That is, how
many squares were there on the grid?
b What is the total energy of the system? That is, what was the total
number of energy quanta or how many counters were used?
c Calculate $\dfrac{n_1}{n_0}, \dfrac{n_2}{n_1}, \dfrac{n_3}{n_2}$ and so on.
What do you notice? What do you think might be the significance of the
result?

14 Consider two molecules, A and B. At the start, A is 'hot' and possesses 7
quanta of energy, and B is 'cool', possessing 2 quanta of energy. Suppose
the two molecules are put together and allowed to exchange quanta.
a Draw up a table like the ones on page 96 to show all the different ways A
and B can share energy.
b What is the total number of ways of sharing?
c In how many of the sharing possibilities does A end up with *less* quanta
than it had at the start?
d What is the possibility that A will get 'cooler' when it is allowed to
exchange quanta with B?

15 A student who had just studied the law of conservation of energy was
heard to say: 'Energy can neither be created nor destroyed. Therefore the

These formulae are confirmed by experiments to find the relative number of moles of atoms of each element present in these compounds.

In ionic compounds of two elements such as those mentioned above, the charge on the ion of each element is taken as the oxidation number of that element. In $CaCl_2$, therefore, the oxidation number of calcium is $+2$ and that of chlorine is -1, and in sodium monoxide, Na_2O, sodium and oxygen have oxidation numbers of $+1$ and -2 respectively.

The use of oxidation number can be extended to molecular compounds in the following way. It is found that in all ionic oxides (excluding peroxides, which contain the O_2^{2-} ion), the oxidation number of oxygen is -2. Suppose we give it that number in the molecular compound CO_2. The oxidation number of carbon in this compound must therefore be $+4$. As a check, we can apply this idea to another compound. Chlorine has the oxidation number -1 in all ionic chlorides. Suppose it is also -1 in the molecular compound tetrachloromethane (carbon tetrachloride), CCl_4. The oxidation number for carbon will therefore again be $+4$.

Extensions of this sort enable one to assign an oxidation number to any element in any compound, once the empirical formula of that compound has been determined experimentally. You can try this yourself, using the formula of the hydrides, chlorides, and oxides of the elements of the first, second, and third periods of the Periodic Table (elements hydrogen to argon). Start by drawing up a table with these headings.

Element	Formula of hydride	Formula of chloride	Formula of oxide

Use the *Book of data* table of properties of inorganic compounds to look up the formulae of the compounds that are new to you.

Next copy the chart given in figure 5.1 in your notebook, marking the axes shown.

To complete it, start with ionic compounds, and mark the oxidation numbers of the elements concerned, with 'o'. Sodium and oxygen are already marked as examples. Now extend the idea to molecular compounds, marking the point with 'x' in this case. Carbon is already given as an example.

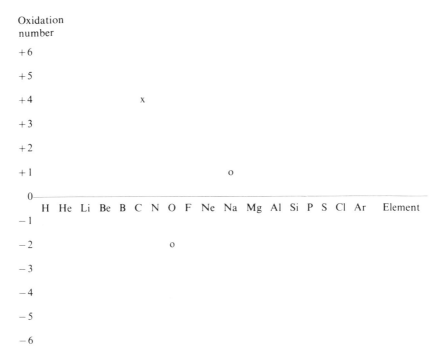

Figure 5.1
Oxidation numbers of the elements H to Ar.

Questions

(Write the answers into your notebook in such a way as to make it quite clear what each question was.)

1 What relationship is there between oxidation number and the group of the Periodic Table in which the element is to be found?

2 What signs are there of a pattern of oxidation numbers along the second period of the Periodic Table being repeated along the third period?

3 What type of element has oxidation numbers which are always positive?

These ideas can be summarized and extended for further use as a set of rules for assigning oxidation numbers. You may find it useful to write abbreviated versions of these rules in your notebook.

is very well shown by constructing an oxidation number chart. The formulae of the various compounds can be written on the chart, as shown in figure 5.2.

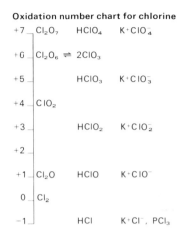

Figure 5.2

Names of the compounds appearing in the chart

In this book, the Stock names for these compounds will be used, but as the old names may still be found, both names are included in the following table. The potassium salts are given as examples; you may care to work out the names of the acids for yourself.

Formula of compound	Oxidation no. of Cl	Stock name	Old name
$KClO_4$	+7	potassium chlorate(VII)	potassium perchlorate
$KClO_3$	+5	potassium chlorate(V)	potassium chlorate
$KClO_2$	+3	potassium chlorate(III)	potassium chlorite
$KClO$	+1	potassium chlorate(I)	potassium hypochlorite
KCl	−1	potassium chloride	potassium chloride

Oxidation and reduction

A change in the oxidation number of an element in a reaction can be used to discover whether the element has been oxidized or reduced. In a particular reaction, a substance which increases the oxidation number of an element is called an oxidizing agent, whereas one which decreases the oxidation number of an element is called a reducing agent. The word 'increases' is taken to mean 'makes more positive or less negative'. Likewise, if a compound high on the oxidation number chart for a given element is to be made from one lower down, an oxidation reaction will be needed, and *vice versa*.

Balancing equations using oxidation numbers

If a reaction involves 'redox' (that is, *red*uction of one element and *ox*idation of another) it is sometimes helpful to use oxidation numbers in balancing the equation for the reaction.

As a very simple example to illustrate the method, consider the reaction between magnesium and hydrochloric acid. Written ionically, the equation involves the following species:

$$Mg(s) + H^+(aq) \longrightarrow Mg^{2+}(aq) + H_2(g)$$

The magnesium initially has oxidation number 0, changing to $+2$ in $Mg^{2+}(aq)$. The hydrogen initially has oxidation number $+1$, changing to 0 in $H_2(g)$. In a reaction, the total change of oxidation number must be the same in both directions, so there must be two $H^+(aq)$ ions reacting with every $Mg^{2+}(aq)$ ion. The balanced equation is therefore:

$$Mg(s) + 2H^+(aq) \longrightarrow Mg^{2+}(aq) + H_2(g)$$

Clearly in such a simple example, the oxidation number method is not really necessary. But if you needed to balance

$$MnO_4^-(aq) + H^+(aq) + Fe^{2+}(aq) \longrightarrow Mn^{2+}(aq) + Fe^{3+}(aq) + H_2O(l)$$

the process would be made very much easier by oxidation numbers.

It is first necessary to identify the elements which actually change in oxidation number. The manganese changes from $+7$ in $MnO_4^-(aq)$ to $+2$ in $Mn^{2+}(aq)$, a change of 5 units. The iron changes from $+2$ in $Fe^{2+}(aq)$ to $+3$ in $Fe^{3+}(aq)$, a change of 1 unit. The total change in oxidation number must be the same in both directions, so there must be five $Fe^{2+}(aq)$ ions reacting with every $MnO_4^-(aq)$ ion. The first stage of balancing thus gives:

Reactions of elements with chlorine

Element examined	Reaction	State of product	Appearance of product	Formula of product
Sodium				
Magnesium				
[etc.]				

EXPERIMENT 5.3b
To investigate the reactions between halogens and halide ions

This experiment investigates the relative reactivity of the halogen elements towards the halide anions.

Use the halogen elements chlorine, bromine, and iodine in solution (in water for the first two of these and in a solution of potassium iodide in water for iodine, as the solubility of iodine in water is small). Fluorine is too hazardous for use under ordinary laboratory conditions.

Handle the solutions with care. Avoid inhaling the vapours from them, and do not allow them to come into contact with your skin or clothing.

Wear eye protection.

Procedure

Set up four test-tubes containing equal volumes of solutions of potassium chloride, potassium bromide, and potassium iodide, and water as a control. Add two or three drops of chlorine solution to each.

Now use colour changes as a guide: have reactions taken place, and what are the products? Would the addition of 1,1,1-trichloroethane help you in reaching a decision?

Now repeat the experiment, using in turn bromine solution and iodine solution. Is a definite trend in reactivity observable in this experiment?

Draw up a table for recording your results, similar to the one that follows.

	Action on			
Solution added	water	potassium chloride solution	potassium bromide solution	potassium iodide solution
Chlorine solution				
[etc.]				

EXPERIMENT 5.3c
To investigate the reactions of the halogens with alkalis

WARNING: The alkalis used in this experiment are very harmful, especially to eyes. Safety glasses MUST be worn, and great care should be taken in doing the experiment. Protective gloves should be worn when clearing up any spillages.

Procedure

Take 2 cm^3 samples of solutions of each of the halogens in water and add a few drops at a time of M sodium hydroxide solution. It should be easy to see what happens to the bromine and iodine because the solutions are coloured; the chlorine is less easy to observe. Record your observations as follows.

Halogen solution	Observations on adding alkali	Equation

Halogens react with cold sodium hydroxide solution according to the pattern set by chlorine:

$$Cl_2 + 2NaOH \longrightarrow NaCl + NaClO + H_2O$$

The compound with formula NaClO is called sodium chlorate(I) or sodium hypochlorite. An ionic equation can be written, leaving out the sodium ions, since these do not undergo chemical change.

Procedure

Titrate $10\,cm^3$ samples of $0.01\,M$ iodine solution with $0.01\,M$ sodium thiosulphate $(Na_2S_2O_3)$ solution. You can measure the iodine solution using a burette or a pipette; if the latter, you must use a pipette filler to fill it. The sodium thiosulphate solution must be delivered from a burette.

You will probably be able to do these titrations without using an indicator because the iodine solution is yellow–red in colour and the products of the reaction are colourless. Nevertheless the end point can be 'sharpened' considerably by adding a few drops of 1% starch when the iodine colour has become very pale. A very dark blue colour is produced which suddenly disappears at the end point of the titration.

Record the details of the experiment in your notebook and give your titration results in the form of a table as in experiment 2.3 on page 29.

Show that your titration results are consistent with the equation for the reaction, which is

$$2Na_2S_2O_3(aq) + I_2(aq) \longrightarrow 2NaI(aq) + Na_2S_4O_6(aq)$$
or ionically,
$$2S_2O_3^{2-}(aq) + I_2(aq) \longrightarrow 2I^-(aq) + S_4O_6^{2-}(aq)$$

Record these equations in your notebook.

Work out the oxidation number of sulphur in sodium thiosulphate, $Na_2S_2O_3$, and in sodium tetrathionate, $Na_2S_4O_6$. It is interesting that the oxidation number of sulphur in sodium tetrathionate contains a fraction. This does not invalidate the use of the oxidation number and the situation is not unusual, particularly in organic chemistry.

This reaction may be used to estimate the concentrations of oxidizing agents which will oxidize iodide ions to iodine. Either or both of the following experiments may now be done.

EXPERIMENT 5.3f
To determine the purity of samples of potassium iodate(v)

Part 1

You are going to find the percentage purity of the potassium iodate(v) from experiment 5.3d. It is quite possible that this contains small amounts of other substances. This experiment is intended to find out how much of a weighed sample of your product is actually potassium iodate(v) and to express this as a percentage.

Procedure

1 Weigh out accurately about 0.05 to 0.1 g of your potassium iodate(v), dissolve it in pure water in a beaker, and transfer the solution through a funnel to a 100 cm³ volumetric flask. Rinse out the beaker several times with water and add the rinsings to the flask. Then make up the volume of the solution to the mark on the neck of the flask with pure water. Mix the contents of the flask well.

2 To 10.0 cm³ portions of this potassium iodate(v) solution, taken with a pipette and pipette filler or with a burette, add about 10 cm³ of approximately 0.1M potassium iodide and about 10 cm³ of M sulphuric acid. The effect of this is to liberate iodine according to the equation:

$$IO_3^-(aq) + 5I^-(aq) + 6H^+(aq) \longrightarrow 3I_2(aq) + 3H_2O(l)$$

3 Titrate each sample with 0.01M sodium thiosulphate, using 1% starch as indicator. Record your results in tabular form.

Calculation

1 How many moles of sodium thiosulphate, $Na_2S_2O_3$, were used in an average titration?

2 How many moles of iodine molecules did these react with? (See the equation in the previous experiment.)

3 How many moles of iodate(v) ions are involved in producing this iodine? (See equation above.)

4 What mass of potassium iodate(v) is this?

5 The mass of pure potassium iodate(v) in 100 cm³ of solution is 10 times this.

6 Calculate the percentage purity of the potassium iodate(v) according to the relationship:

$$\% \text{ purity} = \frac{\text{mass of } KIO_3 \text{ as calculated in } \mathbf{5}}{\text{mass of crude } KIO_3} \times 100$$

Part 2

You are going to find the percentage of potassium iodate(v) remaining in the potassium iodide from experiment 5.3d. The principal impurity in the potassium iodide is likely to be potassium iodate(v). When this mixture is acidified, iodine will be liberated according to the equation:

$$IO_3^-(aq) + 5I^-(aq) + 6H^+(aq) \longrightarrow 3I_2(aq) + 3H_2O(l)$$

Chloride	Effect of adding the chloride to water	Equation
$HCl(g)$	Dissolves readily	$HCl(g) + aq \rightarrow H^+(aq) + Cl^-(aq)$
$LiCl(s)$	Dissolves readily	$LiCl(s) + aq \rightarrow Li^+(aq) + Cl^-(aq)$
$BeCl_2(s)$	Hydrolyses easily	Uncertain, but possibly $BeCl_2(s) + 2H_2O(l)$ $\rightarrow Be(OH)_2(s) + 2H^+(aq) + 2Cl^-(aq)$
$BCl_3(g)$	Hydrolyses violently	$BCl_3(g) + 3H_2O(l)$ $\rightarrow B(OH)_3(aq) + 3H^+(aq) + 3Cl^-(aq)$
$CCl_4(l)$	Immiscible with water	—
$NCl_3(l)$	Hydrolyses	$NCl_3(l) + 3H_2O(l)$ $\rightleftharpoons NH_3(aq) + 3HClO(aq)$
$Cl_2O(g)$	Reacts with water	$Cl_2O(g) + H_2O(l) \rightarrow 2HClO(aq)$
$ClF(g)$	Reacts with water	Uncertain, but possibly $ClF(g) + H_2O(l)$ $\rightarrow H^+(aq) + F^-(aq) + HClO(aq)$
$NaCl(s)$	Dissolves readily	$NaCl(s) + aq \rightarrow Na^+(aq) + Cl^-(aq)$
$MgCl_2(s)$	Dissolves readily with very slight hydrolysis	$MgCl_2(s) + aq \rightarrow Mg^{2+}(aq) + 2Cl^-(aq)$
$AlCl_3(s)$	Hydrolyses	$AlCl_3(s) + 3H_2O(l)$ $\rightarrow Al(OH)_3^*(s) + 3H^+(aq) + 3Cl^-(aq)$
$SiCl_4(l)$	Hydrolyses	$SiCl_4(aq) + 4H_2O(l)$ $\rightarrow SiO_2^*(s) + 4H^+(aq) + 4Cl^-(aq)$
$PCl_3(l)$	Hydrolyses	$PCl_3(l) + 3H_2O(l)$ $\rightarrow H_3PO_3(aq) + 3H^+(aq) + 3Cl^-(aq)$
$S_2Cl_2(l)$	Hydrolyses	$S(s)$, $H^+(aq)$, and $Cl^-(aq)$ are amongst the products
$Cl_2(g)$	Hydrolyses	$Cl_2(g) + H_2O(l)$ $\rightleftharpoons HClO(aq) + H^+(aq) + Cl^-(aq)$

*these are in a hydrated form

5 Broadly, what pattern is there in the behaviour of the chlorides towards water?

WARNING: Many of the liquid and gaseous chlorides in the table are hazardous substances. It is not advisable to carry out reactions involving many of these, even as demonstrations.

There are, however, some properties of halides that can be investigated with relative safety, and they are included in the next experiment.

EXPERIMENT 5.4
To investigate some reactions of the halides

For the first part of this experiment, use solutions of potassium (or sodium) chloride, bromide, iodide, and, if available, fluoride, which are 0.1M with respect to the halide ions. Fluorides are poisonous, so take care in using them. Where you can, attempt to estimate roughly the proportions of the solutions needed for complete reaction.

Procedure

1 To separate 1 cm^3 portions of the halide solutions, add 0.1M silver nitrate solution.

2 To the precipitates obtained in **1** add ammonia solution.

3 Obtain a second set of silver halide precipitates and leave them exposed to the light for an hour.

In parts **4**, **5**, and **6** of this experiment use solid potassium (or sodium) chloride, bromide, and iodide. *Do not use fluoride.* You *MUST* wear eye protection.

4 Investigate the action of concentrated sulphuric acid on the salts. Put about 0.1 g of the solid salt into a test-tube (about enough to fill the rounded end of the tube if it is 100 × 16 mm) and add about 10 drops of concentrated sulphuric acid (*TAKE CARE*). Warm the reaction mixture gently if necessary. Identify as many products as you can, noting the similarities and differences between the reactions. Record and explain your observations as fully as you can.

5 Repeat **4**, using phosphoric acid (*TAKE CARE*) in place of sulphuric acid. Note any difference.

6 Use the reaction in **5** to prepare and collect samples of hydrogen chloride, hydrogen bromide, and hydrogen iodide. The apparatus shown in figure 5.4 is convenient for this purpose. A good yield of gas is obtained if solid 100 per cent phosphoric acid is used. Mix about 2 g halide with an equal quantity of solid phosphoric acid (*TAKE CARE*) in the (side-arm) test-tube. Cork it securely. Put a dry test-tube round the delivery tube and warm the mixture gently until gas is evolved. Collect at least three tubes of gas, corking them when apparently full (when the gas forms copious white fumes at the test-tube mouth). Use the tubes of gas to investigate:

A The solubility of the gas in water. Invert a tube of gas in a beaker of water and remove the cork. If the water rises rapidly the gas is readily soluble. Is there a residue of undissolved gas, and if so, what do you suppose it is?

B The reaction of the gas with ammonia gas. Hold a drop of fairly concentrated ammonia solution in the mouth of an open test-tube, using a glass

very dangerous indeed, and mixtures of the solids with many other substances explode in a violent and unpredictable manner. Because of this it is essential to check the names of the various compounds of the halogens with great care before carrying out any experiments. Remember that if the words chlorate, bromate, or iodate in a name are *not* followed by a Roman numeral, the names are the old names for chlorate(v), bromate(v), and iodate(v).

EXPERIMENT 5.6
Some reactions of the potassium halates(v)

The three potassium halates(v) which are used in this set of experiments have the formulae $KClO_3$, $KBrO_3$, and KIO_3.

Procedure

1 *Action of heat* Using dry, hard-glass test-tubes, investigate the action of heat on each of the solid potassium halates(v), testing for oxygen in each case. Try to compare the three halates(v) in their ease of decomposition.

2 *Oxidation of iron(II) ions*, Fe^{2+}(aq) Make a solution of each of the halates(v) in turn and acidify each with M sulphuric acid. *WARNING:* check that you are using *dilute* acid. Attempt to oxidize samples of iron(II) sulphate solution with each halate(v) and try to explain what you see.

3 *Oxidation of iodide ions*, I^-(aq) Use the acidified samples of each of the halates(v) to attempt to oxidize iodide ions. Try to explain your observations. You have encountered one of these reactions previously in this Topic. Write an account of these reactions and their explanations in your notebook, comparing the behaviour of the chlorates, bromates, and iodates, as far as possible.

Try to write equations for the reactions you have seen in this experiment.

BACKGROUND READING 1
Sources of the halogens

All the halogens are in use commercially, either as the free element or in compounds. They are obtained from a variety of sources and extracted by a number of different methods depending on convenience and cost. Some information is summarized in table 5.2.

Fluorine compounds occur in a number of rocks but these compounds are so widely dispersed that deposits which can be worked economically are

Halogen	Abundance (parts per million by mass) in rocks in the sea	Source	Relative cost of sodium halide (reagent grade)	Chemical process to obtain the free element
Fluorine	700 1.4	fluorite, CaF_2, e.g. Derbyshire 'Blue John'	5	Electrolysis of a solution of potassium fluoride in anhydrous hydrogen fluoride.
Chlorine	200 19 000	rock salt, NaCl, and sea water	1	Electrolysis of an aqueous saturated solution of sodium chloride.
Bromine	3 67	sea water	2	Oxidation of bromide (aq) by chlorine.
Iodine	0.3 0.05	caliche, $NaNO_3$, containing $NaIO_3$	7	Reduction of iodate (aq) by sodium hydrogensulphite.

Table 5.2

rare. By contrast iodine, with much the lowest overall abundance of the halogens, occurs in extensive deposits in the Chilean desert. The principal mineral is caliche, or sodium nitrate. It contains iodine compounds at a concentration of 1500 parts per million, high enough to make the extraction of the element economically feasible.

Sea water is potentially a good source of chemicals, and chlorine and bromine are amongst the elements whose compounds are obtained from it.

The halogens are found in a variety of locations other than those of commercial importance and an explanation of some of these freaks of nature poses difficult questions for geochemists.

The occurrence of hydrogen fluoride and hydrogen chloride gases in nature is surprising, as they are so reactive. They are usually associated with volcanic action. In the Valley of Ten Thousand Smokes in Alaska, for example, over one million tonnes of hydrogen chloride and nearly a quarter of a million tonnes of hydrogen fluoride are emitted every year!

Geochemists are also puzzled by the Chilean desert deposits. On a desolate plateau just inland from the Pacific lies a deposit of soluble salts 30 kilometres wide and stretching for over 500 kilometres (the distance between Leeds and Plymouth) that is quite unique in the World. How did iodine come to be present in such a high concentration? How can the high oxidation number of the elements be explained, nitrogen as nitrate, iodine as iodate, and chromium as chromate? These questions cannot be answered with assurance.

The biologist also finds difficult questions posed by the halogens, for a

supporting structures such as the bones. In humans, there is a high concentration in teeth, especially in the enamel. At the present time there appears to be no known specific role for the fluoride ion in metabolism.

A large number of dental studies have shown that fluoride ions in low concentrations in drinking water can effectively arrest the development of tooth decay in children. As a result of these studies, fluoride ions are now added to the drinking water supply in certain areas in Great Britain. However, the concentration of the fluoride ion must be critically controlled because at slightly higher concentrations the ion causes mottling of the dental enamel, and at much higher concentrations the ion is toxic to life. But at the correct concentration in drinking water it is beneficial to the healthy development of teeth.

Chlorine, like fluorine, is very toxic to living systems and was used as a war gas in the First World War. This halogen is sufficiently soluble in water to be useful as an antibacterial solution and for this reason it is often added to the drinking water at concentrations of 0.1 to 0.5 part per million.

The chloride ion is the principal anion found in the fluid which bathes our body cells (the extracellular fluid); blood plasma forms a significant proportion of this fluid. In this extracellular fluid the chloride ion plays an important role in the maintenance of the osmotic equilibrium between the intracellular fluid and the extracellular fluid. The concentration of chloride ions in the blood plasma is about 365 mg per 100 cm^3 which closely resembles the concentration of chloride in sea water. From this fact attempts have been made to draw conclusions that life originated in the sea.

Apart from the abundance of the chloride ion in blood plasma, it is also present in sweat and saliva. Consequently, during bouts of hard physical activity, or if we have to live in a hot climate, situations in which we would sweat more than usual, it is essential that we increase the intake of salt to compensate for the increased losses of sodium and chloride ions. Muscular cramp is one of the first symptoms of this salt deficiency.

Bromine is also too powerful an oxidizing agent to be encountered in living systems but the bromide anion occurs in small amounts. This ion is readily absorbed from the diet and, unlike the fluoride or the chloride ion, the bromide ion exhibits a highly specific effect on the central nervous system. Bromides depress the higher centres of the brain so that at the correct dosage the effect is one of sedation, but at higher dosage the effect is drowsiness and sleep. Bromides, unlike the fluorides, are not concentrated in any one tissue in the body and are eliminated in much the same way as the chlorides, namely by urinary excretion.

The element iodine is essential for humans and all other mammals, and we derive much of our daily requirement from small amounts of iodide ions which are present in common salt as a trace contaminant. In mammals the iodide is concentrated by a small endocrine gland which is located in the throat and called the thyroid. The iodide trapping mechanism in the thyroid is not fully understood but after the anion is concentrated it is then subjected to an oxidation–reduction reaction and converted from iodide to iodine. The iodine is then involved in a series of reactions which eventually yield the hormone thyroxine. This hormone is then secreted by the thyroid gland and circulates via the blood; it is picked up by almost all cells and tissues. Thyroxine influences the rate of metabolism of body tissues, in particular the rate of oxygen uptake. In certain communities where the drinking water is low in iodide and the diet does not contain any other sources of iodide, there is the possibility of iodine deficiency disease developing. This can be prevented by supplying such communities with common salt to which iodide has been added.

SUMMARY

At the end of this Topic you should:

1 be able to assign oxidation numbers to elements in compounds;
2 be able to balance equations, using oxidation numbers;
3 know some of the reactions of the halogens in oxidation numbers

$-1, 0, +1,$ and $+5$;

4 be able to do titrations involving sodium thiosulphate and their associated calculations;
5 be aware of the sources of the halogens, their manufacture and uses, and their importance in human metabolism.

PROBLEMS

1 Consider the first element in each of the following reactions and state whether its oxidation number goes:

A – up,   B – down, C – remains the same

i $Ag^+(aq) + Cl^-(aq) \longrightarrow AgCl(s)$
ii $Zn(s) + 2H^+(aq) \longrightarrow Zn^{2+}(aq) + H_2(g)$
iii $2Sr(s) + O_2(g) \longrightarrow 2SrO(s)$
iv $2Na(s) + Cl_2(g) \longrightarrow 2NaCl(s)$
v $Cl_2(g) + 2Na(g) \longrightarrow 2NaCl(s)$
vi $Co^{2+}(aq) + \frac{1}{2}Cl_2(g) \longrightarrow Co^{3+}(aq) + Cl^-(aq)$
vii $H_2(g) + Cl_2(g) + aq \longrightarrow 2HCl(aq)$
viii $I^-(aq) + \frac{1}{2}Br_2(aq) \longrightarrow \frac{1}{2}I_2(aq) + Br^-(aq)$
ix $Cl^-(aq)$ (at anode) $\longrightarrow \frac{1}{2}Cl_2(g) + e^-$
x $\frac{1}{2}O_2(g) + H_2(g) \longrightarrow H_2O(g)$
xi $BaCl_2(s) + aq \longrightarrow Ba^{2+}(aq) + 2Cl^-(aq)$

TOPIC 6
Energy changes and bonding

In nearly all chemical reactions – whether in the laboratory when, for example, solutions of an acid and an alkali are mixed, or in a power station when coal or oil is burning – there is an energy change. The study of these energy changes is as much a part of chemistry as the study of the changes of materials in a reaction or of the structure of substances.

In this Topic we shall investigate these energy changes, and in measuring them we shall seek to answer two questions in particular.

1 It is sometimes possible to convert a substance A into another substance B by several different routes. Does the route by which a chemical change takes place make any difference to the overall energy change?

2 When a chemical reaction occurs, there is a change in the nature and perhaps the number of bonds between atoms. Can we say that particular bonds make specific contributions to overall energy changes?

The answers to these questions will lead us to important conclusions about energy changes. These in turn will help to increase our understanding of the reasons why chemical reactions take place.

6.1
ENERGY CHANGES – DEFINITIONS

If we are to investigate energy changes, we must state the conditions under which the changes are measured, so that the results can be compared.

The energy change that we shall be measuring in this Topic is the enthalpy change. This can be considered to be the heat that would be exchanged with the surroundings if the reaction occurred in such a way that the temperature and pressure of the system before and after the reaction were the same.

The enthalpy change of a reaction is the heat exchange with the surroundings *at constant pressure*.

heat exchange with the surroundings

reactants at temperature T	$\longrightarrow$	products at temperature T
and pressure p with		and pressure p with
enthalpy $= H_1$		enthalpy $= H_2$

The symbol Δ (Greek capital delta) is used to denote the change in the value of a physical quantity. The change in enthalpy going from reactants to products, ΔH, is given by

$$\Delta H = H_2 - H_1$$

Normally we insulate the system from its surroundings, and allow the heat of the reaction to change the temperature of the system. We then calculate how much heat would have to be put into or taken from the system to bring it back to its initial temperature. This amount of heat is the enthalpy change.

The enthalpy change is not the only energy change that could be measured. Suppose a reaction in which a gas is produced is carried out at constant pressure. The reaction between zinc and dilute sulphuric acid is an example

$$Zn(s) + 2H^+(aq) \longrightarrow Zn^{2+}(aq) + H_2(g)$$

As the hydrogen is made, the system will have to expand against the atmosphere and work will have to be done by the system on the surroundings. The enthalpy change includes this energy as well as the energy change due to changes in bonding. If the reaction is conducted at constant volume, this work does not have to be done, and the energy change under these conditions is called the internal energy change, and given the symbol ΔU. (See figure 6.1.)

As it is more convenient to carry out reactions in the laboratory under conditions of constant pressure (in open beakers and test-tubes), we normally refer to enthalpy changes rather than internal energy changes. In reactions in which there is no volume change at constant pressure, enthalpy change = internal energy change. When there is a volume change, the difference is less than 5 per cent for reactions in which the value of ΔH is greater than $40 \, kJ$.

If the reaction is exothermic, that is, if heat is given out from the system to the surroundings during the reaction, then the enthalpy of the reactants, H_1, must be greater than that of the products, H_2, so the enthalpy change

$$\Delta H = H_2 - H_1 \text{ is negative.}$$

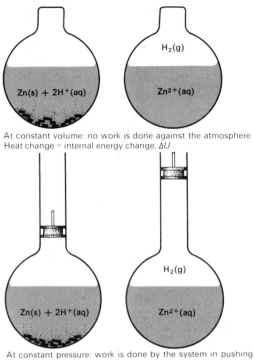

At constant volume: no work is done against the atmosphere
Heat change = internal energy change, ΔU

At constant pressure: work is done by the system in pushing
the atmosphere back   Heat change = enthalpy change, ΔH

Figure 6.1
Reactions at constant volume and pressure.

Conversely if the reaction is endothermic, that is, if heat is taken into the system from the surroundings during the reaction, then H_2 must be greater than H_1 and the enthalpy change

$$\Delta H = H_2 - H_1 \text{ is positive.}$$

For the enthalpy change

$$\text{C (graphite)} + O_2(g) \longrightarrow CO_2(g); \qquad \Delta H = -393.5 \text{ kJ mol}^{-1}$$

the value of ΔH indicated is for the amounts shown in the equation, that is, for one mole of carbon atoms, C, one mole of oxygen molecules, O_2, and one mole of carbon dioxide molecules, CO_2. Normally, a standard enthalpy change is quoted which refers to that at 101 kPa (1 atmosphere) pressure and at some particular temperature, and it is then given the symbol $\Delta H^{\ominus}$. The superscript $^{\ominus}$ indicates standard at some temperature which must be stated. The normal temperature used is 298 K and this is shown as a subscript to the symbol, thus:

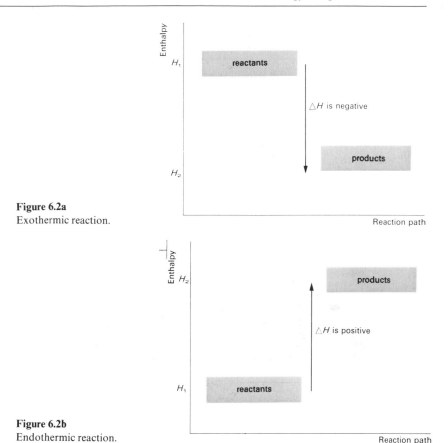

Figure 6.2a
Exothermic reaction.

Figure 6.2b
Endothermic reaction.

$\Delta H^{\ominus}_{298}$. This symbol also means that the substances must be in the physical state normal at 298 K and 101 kPa, that is, solid carbon*, gaseous oxygen, and gaseous carbon dioxide. If there are any solutions involved, then the standard condition is a concentration of one mole per cubic decimetre (1M).

It is not possible to find the standard enthalpy change of formation of, say, carbon dioxide, at 298 K directly, because carbon and oxygen do not react at this temperature. However, it is possible to determine a value at a temperature at which they do react, and calculate a value at 298 K from this result.

The standard enthalpy change for a reaction, symbol $\Delta H^{\ominus}_{298}$, refers to the amounts shown in the equation, at a pressure of 101 kPa, at a temperature of 298 K, with the substances in the physical states normal under these conditions. Solutions must have a concentration of 1 mol dm^{-3}.

* In the case of elements or compounds which can exist in different forms such as carbon, the most stable form is chosen as the standard. In this case the most stable form is graphite.

For certain reactions we have a shorthand which saves us from having to write the full equation. The reaction quoted above is one such; the enthalpy change for it refers to the formation of one mole of carbon dioxide molecules. It is given the symbol $\Delta H^{\ominus}_{f,298}$ [$CO_2(g)$], and it is called the standard enthalpy change of formation of carbon dioxide.

The standard enthalpy change of formation of a compound, symbol $\Delta H^{\ominus}_{f,298}$ is the enthalpy change that takes place when one mole of the compound is formed from its elements under the standard conditions.

It follows from this definition that the standard enthalpy change of formation of an element in its standard state is zero, for example:

$$Hg(l) \longrightarrow Hg(l); \qquad \Delta H^{\ominus}_{f,298}[Hg(l)] = 0$$

However, if the state of the element is changed, as for example

$$Hg(l) \longrightarrow Hg(g); \qquad \Delta H^{\ominus}_{f,298}[Hg(g)] = +59.15\,kJ\,mol^{-1}$$

Another energy change that is specially defined is the standard enthalpy change of atomization of an element, $\Delta H^{\ominus}_{at,298}$. This refers to the enthalpy change when one mole of gaseous atoms is formed from the element in the defined physical state.

$$\tfrac{1}{2}H_2(g) \longrightarrow H(g); \qquad \Delta H^{\ominus}_{at,298}[H_2(g)] = +218\,kJ\,mol^{-1}$$

The standard enthalpy change of atomization of an element, symbol $\Delta H^{\ominus}_{at,298}$, is the enthalpy change that takes place when one mole of gaseous atoms is made from the element in the defined physical state under standard conditions.

Other reactions for which a special definition applies are those involving combustion.

The standard enthalpy change of combustion of a substance, symbol $\Delta H^{\ominus}_{c,298}$, is defined as the enthalpy change that occurs when one mole of the substance undergoes complete combustion under standard conditions.

For a compound containing carbon, for example, complete combustion means the conversion of the whole of the carbon to carbon dioxide, as shown in the following equation.

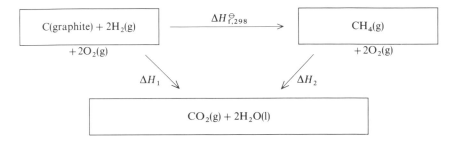

ΔH_1 is made up of the standard enthalpy change of combustion of graphite, and twice the standard enthalpy change of combustion of hydrogen.

$$C(graphite) + O_2(g) \longrightarrow CO_2(g); \quad \Delta H^{\ominus}_{c,298}[C(graphite)] = -393.5\,kJ$$
$$2H_2(g) + O_2(g) \longrightarrow 2H_2O(l); \quad 2 \times \Delta H^{\ominus}_{c,298}[H_2(g)]$$
$$= 2 \times (-285.8)\,kJ$$
$$= -571.6\,kJ$$

Therefore

$$\Delta H_1 = -393.5 + (-571.6)\,kJ$$
$$= -965.1\,kJ$$

ΔH_2 is the standard enthalpy change of combustion of methane

$$\Delta H_2 = \Delta H^{\ominus}_{c,298}[CH_4(g)] = -890.3\,kJ\,mol^{-1}$$

We may now substitute these values in the earlier diagram, to give

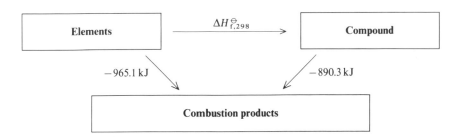

The overall energy change in going from elements to combustion products must be the same whatever the route. Equating the two routes,

$$-965.1 = \Delta H^{\ominus}_{f,298} - 890.3$$
$$\Delta H^{\ominus}_{f,298} = -74.8\,kJ\,mol^{-1}$$

chloride. The standard enthalpy changes of formation are

$$\Delta H^{\ominus}_{f,298}[NH_3(g)] = -46\,\text{kJ mol}^{-1}$$
$$\Delta H^{\ominus}_{f,298}[HCl(g)] = -92\,\text{kJ mol}^{-1}$$
$$\Delta H^{\ominus}_{f,298}[NH_4Cl(s)] = -314\,\text{kJ mol}^{-1}$$

$\Delta H^{\ominus}_{298}$ for the reaction can be calculated as follows. First write down the equation for the reaction:

$$NH_3(g) + HCl(g) \longrightarrow NH_4Cl(s)$$

Then draw a diagram showing the formation of the compounds on both sides of the equation from the same elements:

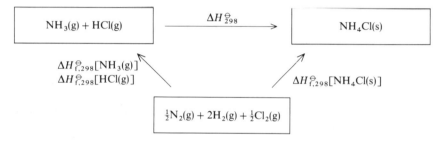

The total enthalpy change must be the same by whatever route the ammonium chloride is formed (whether it is formed 'direct' from its elements, or through the intermediates of ammonia and hydrogen chloride).
Therefore

$$\Delta H^{\ominus}_{f,298}[NH_4Cl(s)] = \Delta H^{\ominus}_{f,298}[NH_3(g)] + \Delta H^{\ominus}_{f,298}[HCl(g)] + \Delta H^{\ominus}_{298}$$

That is

$$-314 = -46 - 92 + \Delta H^{\ominus}_{298}$$

So

$$\Delta H^{\ominus}_{298} = 46 + 92 - 314 = -176\,\text{kJ mol}^{-1}$$

This example shows that standard enthalpy changes of chemical reactions can be calculated from the standard enthalpy changes of formation of the reactants and products, without having to construct Hess's Law cycles for each individual case. This is the great value of standard enthalpy changes of formation;

they systematize enthalpy change calculations, and make it possible to calculate enthalpy changes which cannot otherwise be found.

A useful standard enthalpy change of formation for many chemical reactions is that for ions in aqueous solution. The *Book of data* contains a table of these values. They are used in calculating enthalpy changes of reactions in aqueous solution in exactly the same way as any other standard enthalpy change of formation, as shown in the following example.

What is the enthalpy change for the reaction:

$$Cl_2(g) + 2Br^-(aq) \longrightarrow Br_2(l) + 2Cl^-(aq)?$$

Solution

The standard enthalpy changes of formation of elements are zero, so

$$\begin{aligned}
\Delta H_{reaction} &= 2 \times \Delta H^{\ominus}_{f,298}[Cl^-(aq)] - 2 \times \Delta H^{\ominus}_{f,298}[Br^-(aq)] \\
&= -334.2 - (-242.8) \\
&= -91.4 \text{ kJ mol}^{-1}
\end{aligned}$$

6.4
BOND ENERGIES

Can enthalpy changes of combustion give information about the energy required to break individual bonds?

One way of attempting to answer this question would be to find a series of substances which are closely related to each other, and which differ from each other by some fixed unit of structure. Then by studying such substances it might be possible to see whether that fixed unit of structure makes any consistent contribution to the overall energy situation.

An example would be the series of alcohols:

$CH_3CH_2CH_2OH$	Propan-1-ol
$CH_3CH_2CH_2CH_2OH$	Butan-1-ol
$CH_3CH_2CH_2CH_2CH_2OH$	Pentan-1-ol
$CH_3CH_2CH_2CH_2CH_2CH_2OH$	Hexan-1-ol
$CH_3CH_2CH_2CH_2CH_2CH_2CH_2OH$	Heptan-1-ol
$CH_3CH_2CH_2CH_2CH_2CH_2CH_2CH_2OH$	Octan-1-ol

Each compound differs from the rest by one $-CH_2-$ unit. This situation will be made very clear if you examine structural models of these compounds.

In this series, the question can be posed 'does the $-CH_2-$ group make a specific contribution to the enthalpy change of combustion of alcohols?'. One method of finding out would be to burn the alcohols and measure the enthalpy change per mole of each.

EXPERIMENT 6.4
To find the enthalpy changes of combustion of some alcohols

In this experiment, you will find the enthalpy change of combustion of one member of the series of alcohols and compare your results with those obtained by other members of the class.

Procedure

Use the combustion calorimeter with which you have been provided.

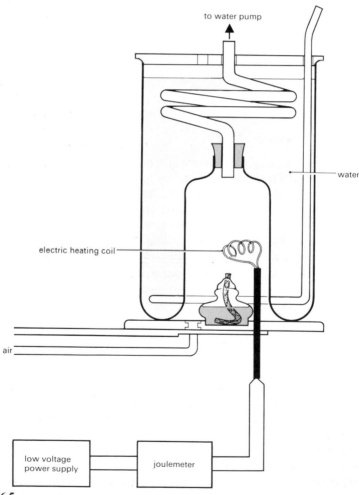

Figure 6.5

1 Put water in the calorimeter up to the level shown and mount it on its stand. Attach a water pump on the top of the copper spiral outlet, and adjust the pump so as to draw a moderately rapid stream of air through the spiral.

2 Almost fill the small spirit lamp with the alcohol to be used, then light the wick, and adjust the length so as to give a flame about 1.5–2 cm high. Put the lamp under the calorimeter and watch its behaviour. If the flame remains reasonably steady but slowly goes out this probably indicates that not enough air is being supplied; adjust the water pump so that more air is drawn through. If the flame is very unsteady and goes out, it may indicate that too great a rush of air is being drawn in; adjust the water pump accordingly.

3 When you have adjusted the height of the wick and the flow of air so that the burner remains alight with a good flame, extinguish the burner and put the cap over the wick. Weigh it on a balance reading to 0.001 g.

4 Stir the water in the calorimeter and record the temperature, using a thermometer reading to 0.1 °C.

5 Remove the cap from the spirit lamp, light the wick, and without delay put the lamp under the calorimeter. Stir the calorimeter periodically. When a rise in temperature of between 10 °C and 11 °C has been obtained blow out the flame, remove the lamp from the stand, and replace the cap. Stir the water thoroughly and note the maximum temperature that is reached.

6 Reweigh the spirit lamp as soon as possible after extinguishing it.

7 Supply electrical energy to the calorimeter via the heater until a similar temperature rise to that produced by the burning alcohol is obtained.

Calculation

Calculate the moles of alcohol molecules burned and hence ΔH_c, the enthalpy change of combustion, in kilojoules per mole.

A consideration of sources of error

1 *Heat losses* Do you think the method has taken heat losses from the calorimeter into account satisfactorily?

2 *Combustion products* The term 'standard enthalpy change of combustion' implies complete combustion, in this instance to carbon dioxide and water. Can we be sure that this has taken place? What might have been formed instead, to some extent?

Do you think that the various sources of error will tend to make your value higher or lower than the true one?

Compare your results with those of others in the class, and see whether they provide any answer to the question with which this section began.

Plot a graph of ΔH_c for the alcohols from methanol to octan-1-ol against the number of carbon atoms. Use your own values and those of other members of your group or take values from the *Book of data*. You will have found that the difference in value between successive alcohols in the series is about the same; and, of course, the structural difference between successive alcohols is the CH_2 group of atoms. When one extra CH_2 group burns, extra energy must be supplied to break one extra C—C bond and two extra C—H bonds; and extra energy is released by the formation of one extra C—O bond and two extra O—H bonds. If a fixed amount of energy is associated with this number of bonds broken and bonds formed, as is found experimentally, it seems likely that each individual bond has its own energy that must be supplied to break it, or that will be released when it is formed.

Bond energies in other compounds

Consider now the alkane series of hydrocarbons and begin with the first member, methane, CH_4. It seems reasonable to assume that the energy associated with the C—H bonds must be reflected in the total amount of energy required to break the molecule into its constituent atoms.

This total amount of energy can be found using Hess's Law. The equation is $CH_4(g) \longrightarrow C(g) + 4H(g)$.

The enthalpy changes are shown on the diagram

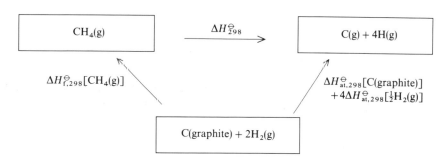

So

$$\Delta H_{298}^{\ominus} + \Delta H_{f,298}^{\ominus}[CH_4(g)] = \Delta H_{at,298}^{\ominus}[C(graphite)] + 4\Delta H_{at,298}^{\ominus}[\tfrac{1}{2}H_2(g)]$$

Putting in the values

$$\Delta H_{298}^{\ominus} = -(-74.8) + 716.7 + (4 \times 218)$$
$$= +1663.5 \, kJ \, mol^{-1}$$

So, for the reaction

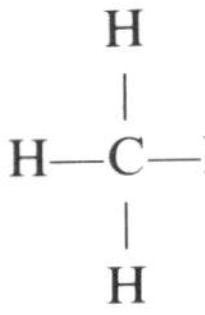

H
|
H—C—H(g)⟶ C(g) + 4H(g); $\Delta H^{\ominus}_{298} = +1663.5\,\text{kJ mol}^{-1}$
|
H

If the bonds are equal in strength, then the bond energy of one C—H bond should be $+\dfrac{1663.5}{4} = +415.9\,\text{kJ mol}^{-1}$.

Denoting the bond energy of the C—H bond by $E(\text{C—H})$ we have $E(\text{C—H}) = +415.9\,\text{kJ mol}^{-1}$.

Now consider ethane, C_2H_6. A similar calculation to the one above shows that for the reaction

$C_2H_6(g) \longrightarrow 2C(g) + 6H(g)$; $\Delta H^{\ominus}_{298} = +2826.1\,\text{kJ mol}^{-1}$

This reaction involves the breaking of six C—H bonds and one C—C bond. Denoting the bond energy of the C—C bond by $E(\text{C—C})$, we have

$+2826.1 = E(\text{C—C}) + 6E(\text{C—H})$

Substituting the value 415.9 for $E(\text{C—H})$ we have

$+2826.1 = E(\text{C—C}) + (6 \times 415.9)$

So $E(\text{C—C}) = +330.7\,\text{kJ mol}^{-1}$

The bond energy $E(\text{C—Cl})$ has been determined, using several compounds. Below are shown the compounds and the values obtained for them:

Compound		$E(\text{C—Cl})$ /kJ mol^{-1}
Cl | Cl—C—Cl | Cl	tetrachloromethane	+ 327
H | H—C—Cl | H	chloromethane	+ 335
H H | | H—C—C—Cl | | H H	chloroethane	+ 342

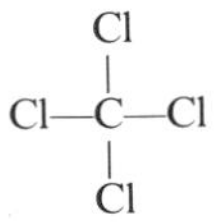

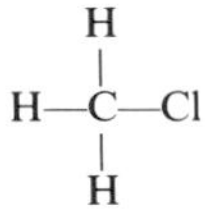

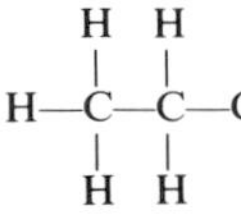

From these examples it can be seen that the bond energy value is approximately the same in each case, though it depends upon the compound from which it was determined, to some extent; that is, the environment of the bond affects the value. The X—Y bond energy will vary somewhat, depending upon the nature of the other atoms or groups of atoms which are attached to X and Y.

But if an average bond energy is taken this can often be very useful. Tables have therefore been prepared giving average bond energies. Average bond energies per mole of bonds are denoted by the symbol E. Some examples are:

Bond	E
	$/kJ\,mol^{-1}$
C—H	413
C—C	347
C—O	358
C—Cl	346
O—H	464

A fuller table is given in the *Book of data*.

An approximate value for the enthalpy change involved in the atomization of a compound from the gaseous state can be obtained by adding up the average bond energies for all the bonds in the molecule of that compound.

When you have read this explanation of bond energies, carry out the following exercises.

1 Using the table of bond energies given in the *Book of data*, work out an approximate value for the energy needed to atomize one mole of the alcohol propan-1-ol, $CH_3CH_2CH_2OH$.

2 Make a table showing the bond energies for the hydrides across the Periodic Table, C—H, N—H, O—H, and F—H, and then insert the vertical series F—H, Cl—H, Br—H, and I—H. What are the trends in the ease of breaking the bonds, and what information can you deduce from them?

BACKGROUND READING 1
Accurate experimental thermochemistry

Now that you have carried out some thermochemical determinations yourself, you may be interested to read how such measurements can be carried out very accurately.

Figure 6.6a shows a modern bomb calorimeter, designed to determine accurately the energy changes that take place when substances are completely burned in oxygen. A crucible, containing a weighed quantity of the substance under investigation, is put in a stainless steel bomb capable of withstanding high pressures. A cross-section of such a bomb is shown in figure 6.6b. The

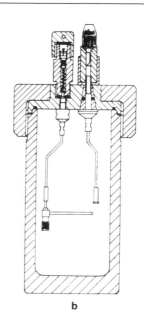

a   b

Figure 6.6
a A modern automatic bomb calorimeter.
b A sectional drawing of the bomb, showing the electrodes which pass through the cap.
A. Gallenkamp & Co. Ltd.

bomb is closed, and filled with oxygen under pressure. It is then put in the calorimeter vessel which has previously been filled with water. The lid of the vessel is lowered, placing in position a stirrer, a thermometer, and connections to the electrodes which pass through the cap of the bomb. When all is ready, the initial temperature is recorded, the substance is ignited electrically, and after a suitable period the final temperature is noted. Corrections for cooling are eliminated as an automatic heating system prevents exchange of heat between the bomb and the surrounding water, by ensuring that the water is kept at the same temperature as the bomb throughout the test.

Bomb calorimeters are usually calibrated using benzoic acid, a compound readily obtainable in a very pure form, whose energy change on combustion is known accurately. If a weighed quantity of benzoic acid is completely burned in oxygen in the bomb, the number of joules required to raise the temperature of the calorimeter system by 1 °C can be found. This value can then be used to convert temperature rises produced by other combustion experiments into the number of joules that caused them.

It should be noted that the energy changes are determined at *constant volume*, and are not therefore enthalpy changes; the enthalpy changes can, however, be calculated from the results obtained.

6.5
THE BORN–HABER CYCLE: LATTICE ENERGIES

Just as it is often useful to know the enthalpy change of formation of a molecular compound from atoms in the gaseous state, so also it is often useful to know the enthalpy change of formation of an ionic crystal from ions in the gaseous state. This latter quantity is often called the 'lattice energy' of the compound.

The *lattice energy* of an ionic crystal is the standard enthalpy change of formation of the crystal lattice from its constituent ions in the gas phase.*

$$Na^+(g) + Cl^-(g) \longrightarrow Na^+Cl^-(s); \quad \Delta H^{\ominus}_{298} = \text{lattice energy}$$

The direct determination of lattice energies is not possible, but values can be obtained indirectly by means of an energy cycle, known as a Born–Haber cycle.

The cycle is analogous with that for obtaining the enthalpy change of formation of molecules from atoms, and may be seen as a triangular two-route process.

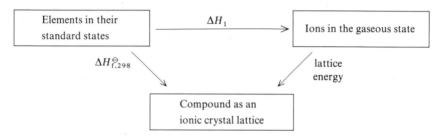

In the case of sodium chloride this is:

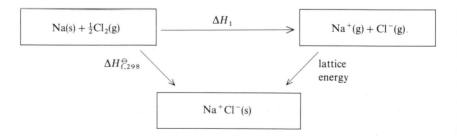

The standard enthalpy change of formation of sodium chloride can be measured directly, by the reaction of sodium with chlorine in a calorimeter. If

*Strictly speaking, this quantity is the lattice enthalpy, ΔH_1. The true lattice energy, U, is related to the lattice enthalpy by the equation $\Delta H_1 = U + \Delta nRT$ as explained in the *Book of data*.

the energy required to convert sodium metal into gaseous ions, and chlorine molecules into gaseous ions, can be obtained, ΔH_1 will be known, and it is then possible to obtain a value for the lattice energy.

ΔH_1 has to be obtained in stages. Taking the sodium first,

$$Na(s) \xrightarrow[\substack{\text{standard enthalpy} \\ \text{change of atomization}}]{} Na(g) \xrightarrow[\substack{\text{ionization} \\ \text{energy}}]{-e^-} Na^+(g)$$

The two energy values required are the *standard enthalpy change of atomization* of sodium, for the conversion of solid sodium into gaseous sodium consisting of separate atoms:

$$Na(s) \longrightarrow Na(g); \qquad \Delta H^{\ominus}_{at,298} = +107.3 \text{ kJ mol}^{-1}$$

and the *ionization energy*, for the conversion of gaseous atoms into gaseous ions:

$$Na(g) \longrightarrow Na^+(g) + e^-; \qquad \Delta H^{\ominus}_{i,298} = +496 \text{ kJ mol}^{-1}$$

Taking the chlorine we have

$$\tfrac{1}{2}Cl_2(g) \xrightarrow[\substack{\text{standard enthalpy} \\ \text{change of atomization}}]{} Cl(g) \xrightarrow[\substack{\text{electron} \\ \text{affinity}}]{+e^-} Cl^-(g)$$

The two energy values required are the *standard enthalpy change of atomization* of chlorine, for the conversion of gaseous chlorine molecules into gaseous chlorine atoms,

$$\tfrac{1}{2}Cl_2(g) \longrightarrow Cl(g); \qquad \Delta H^{\ominus}_{at,298} = +121.7 \text{ kJ mol}^{-1}$$

and the *electron affinity*, which is the energy change occurring when a chlorine atom accepts an electron and becomes a chloride ion,

$$Cl(g) + e^- \longrightarrow Cl^-(g); \qquad \Delta H^{\ominus}_{e,298} = -348.8 \text{ kJ mol}^{-1}$$

Each of these can be determined experimentally, although the determination of electron affinity is difficult.

The only other value to place in the cycle is $\Delta H^{\ominus}_{f,298}[Na^+Cl^-(s)]$:

$$Na(s) + \tfrac{1}{2}Cl_2(g) \longrightarrow Na^+Cl^-(s); \qquad \Delta H^{\ominus}_{f,298} = -411.2 \text{ kJ mol}^{-1}$$

and then the lattice energy can be determined.

From figure 6.7 it can be seen that the lattice energy is

$$-[(121.7 + 496 + 107.3 + 411.2) - 348.8] \text{ kJ mol}^{-1} = -787.4 \text{ kJ mol}^{-1}$$

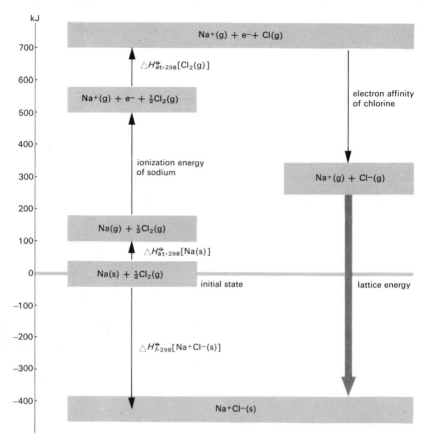

Figure 6.7
Energy level diagram for the formation of sodium chloride.

Theoretical values for lattice energies

A lattice energy is the energy change involved in bringing well separated electrostatic charges together to form a lattice. It should therefore be possible to make an estimate of the magnitude of this energy change, using the principles of electrostatics. The calculations are done on the basis of ions being charged spheres in contact. It is assumed that the ions are spherical, separate entities, each with its charge distributed uniformly around it. Some of the values that have been calculated are given in table 6.1 in the column headed 'Theoretical value'.

Examine values for the alkali metal halides, and compare the theoretical values with the Born–Haber experimental values. For one or two, calculate the approximate percentage discrepancy between theoretical and experimental values.

Compound	Theoretical value	Experimental value (Born–Haber cycle)
NaCl	−770	−780
NaBr	−735	−742
NaI	−687	−705
KCl	−702	−711
KBr	−674	−679
KI	−636	−651
AgCl	−833	−905
AgBr	−816	−891
AgI	−778	−889
ZnS	−3427	−3615

Table 6.1
Lattice energies/kJ mol^{-1}.

The excellent agreement between the theoretical and experimental values is strong evidence that the simple model of an ionic crystal is a good one, in the instances of the alkali metal halides.

Now examine the corresponding values for the silver halides. Calculate the approximate percentage difference between the theoretical and the experimental values. Do you think that the ionic model accurately represents the bonding situation in the silver halides? If not, some other model is required.

Bonding will be discussed again later, in Topic 7.

Lattice energy and stoicheiometry

Can energy considerations give any indication of an expected formula for a compound?

Would energy calculations be able to show, for instance, which of the three formulae, $MgCl$, $MgCl_2$, and $MgCl_3$ would be the most likely for magnesium chloride?

It would be expected that the compound which has the most negative standard enthalpy change of formation would be the most stable. If it is assumed that Mg^+Cl^- would have a sodium chloride lattice structure, and $Mg^{3+}(Cl^-)_3$ a structure similar to $AlCl_3$, then a reasonable estimate of the lattice energies for the hypothetical crystals $MgCl$ and $MgCl_3$ may be made. Born–Haber cycles can then be constructed, and values obtained for the standard enthalpy changes of formation of these hypothetical crystals.

The quantities necessary for drawing the cycles are as follows.

$\Delta H_{298}^{\ominus}/\text{kJ}$

AMg = enthalpy change of atomization of magnesium
  $Mg(s) \rightarrow Mg(g)$ +148
IE$_1$ = 1st ionization energy of magnesium,
  $Mg(g) \rightarrow Mg^+(g) + e^-$ +738
IE$_2$ = 2nd ionization energy of magnesium,
  $Mg^+(g) \rightarrow Mg^{2+}(g) + e^-$. +1451
IE$_3$ = 3rd ionization energy of magnesium,
  $Mg^{2+}(g) \rightarrow Mg^{3+}(g) + e^-$ +7733
ACl = enthalpy change of atomization of chlorine,
  $\frac{1}{2}Cl_2(g) \rightarrow Cl(g)$ +122
2ACl = 2 × enthalpy change of atomization of chlorine,
  $Cl_2(g) \rightarrow 2Cl(g)$ +244
3ACl = 3 × enthalpy change of atomization of chlorine,
  $1\frac{1}{2}Cl_2(g) \rightarrow 3Cl(g)$ +366
 EA = electron affinity of chlorine, $Cl(g) + e^- \rightarrow Cl^-(g)$ −349
2EA = 2 × electron affinity of chlorine, $2Cl(g) + 2e^- \rightarrow 2Cl^-(g)$ −698
3EA = 3 × electron affinity of chlorine, $3Cl(g) + 3e^- \rightarrow 3Cl^-(g)$ −1047
LE$_1$ = estimated lattice energy for MgCl, approximately   −753
LE$_2$ = lattice energy for MgCl$_2$ −2526
LE$_3$ = estimated lattice energy for MgCl$_3$, approximately −5440

The stages involved in the first two processes are:

MgCl

$$Mg(s) + \tfrac{1}{2}Cl_2(g) \xrightarrow{\text{AMg}} Mg(g) + \tfrac{1}{2}Cl_2(g) \xrightarrow{\text{IE}_1} Mg^+(g) + e^- + \tfrac{1}{2}Cl_2(g)$$

$$\downarrow \text{ACl}$$

$$Mg^+Cl^-(s) \xleftarrow{\text{LE}_1} Mg^+(g) + Cl^-(g) \xleftarrow{\text{EA}} Mg^+(g) + e^- + Cl(g)$$

MgCl$_2$

$$Mg(s) + Cl_2(g) \xrightarrow{\text{AMg}} Mg(g) + Cl_2(g) \xrightarrow{\text{IE}_1} Mg^+(g) + e^- + Cl_2(g)$$

$$\downarrow \text{IE}_2$$

$$Mg^{2+}(g) + 2e^- + 2Cl(g) \xleftarrow{\text{2ACl}} Mg^{2+}(g) + 2e^- + Cl_2(g)$$

$$\downarrow \text{2EA}$$

$$Mg^{2+}(g) + 2Cl^-(g) \xrightarrow{\text{LE}_2} Mg^{2+}(Cl^-)_2(s)$$

$MgCl_3$

Construct the sequence for $MgCl_3$ yourself.
The cycles are represented to scale in figure 6.8.

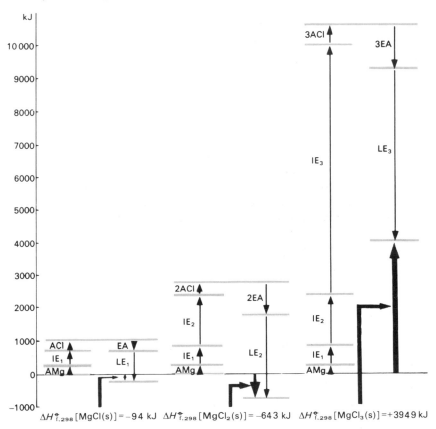

$\Delta H^{\ominus}_{f,298}[MgCl(s)] = -94\ kJ$ $\Delta H^{\ominus}_{f,298}[MgCl_2(s)] = -643\ kJ$ $\Delta H^{\ominus}_{f,298}[MgCl_3(s)] = +3949\ kJ$

Figure 6.8
Born–Haber cycles for chlorides of magnesium.

If you add up the component quantities for the various stages, you will find that the standard enthalpy changes of formation for the compounds, as obtained from the cycles, are:

$$\Delta H^{\ominus}_{f,298}[MgCl(s)] = -94\ kJ\,mol^{-1}$$
$$\Delta H^{\ominus}_{f,298}[MgCl_2(s)] = -643\ kJ\,mol^{-1}$$
$$\Delta H^{\ominus}_{f,298}[MgCl_3(s)] = +3949\ kJ\,mol^{-1}$$

From these values it can be seen that the formation of MgCl is just exothermic;

the formation of $MgCl_2$ is much more exothermic; but the formation of $MgCl_3$ is highly endothermic. It therefore appears that the compound which is formed is the one whose formation involves the greatest transfer of energy to the surroundings.

If the cycles are examined in the scale diagram, it will be seen that the largest single contributions in each cycle are made by the ionization energies and the lattice energy, and that these two are always opposite in sign. Broadly speaking, therefore, the magnitude of the standard enthalpy change of formation depends upon the result of competition between ionization energies and lattice energy. If it requires more energy to ionize the metal than is returned as lattice energy, then the compound will not be formed.

Calculations of this type can be done for many other classes of compound. For instance it may be shown that the formation of NaO would not be energetically favoured, while the formation of Na_2O would be energetically favoured. From your study of the alkali metal oxides you will recall that Na_2O exists while NaO does not.

Thus, energetic considerations help us to understand the stoicheiometry of compounds.

6.6
ENTHALPY CHANGES AND ENTROPY

In Topic 3 we looked at the method of calculating the number of arrangements of molecules in various situations. We saw that 'molecules don't care', and that they always end up in the largest number of ways, W. This principle accounted for the observed behaviour of gases, for example on diffusion.

In Topic 4 we did much the same thing with quanta of energy. The corresponding 'energy doesn't care' principle was seen to account for the observed behaviour of heat, for example in travelling from hot objects to cold ones, because energy too ends up in the condition that has the largest value of W. We introduced the idea of entropy, symbol S, and defined it as

$$S = k \ln W$$

where

k is Boltzmann's constant, and
$\ln W$ is the logarithm of the number of possible arrangements of molecules and quanta of energy.

For a particular physical or chemical situation, the greater the number of possible arrangements, W, the more likely the situation is to come about. For any physical or chemical change which happens spontaneously, of its own accord, W must increase. This in turn means that **entropy must always increase in spontaneous changes**. Another way of putting this is to say that for any spontaneous change, the entropy change, ΔS, must be positive.

Figure 6.8a
The memorial to the great Austrian scientist Ludwig Boltzmann, over his grave in Vienna. It was Boltzmann who first saw the link between the macroscopic laws of thermodynamics and the underlying behaviour of atoms and molecules. Boltzmann devised the equation $S = k \ln W$, whose constant k bears his name. Boltzmann's work was rejected by most other scientists of the time and in 1906 he committed suicide, a disillusioned man.

Photograph, The Boltzmann Society, Vienna.

But there are plenty of examples of spontaneous changes for which ΔS *seems to be negative*. Look at these three examples:

1 *The reaction of magnesium with oxygen*   Once ignited, magnesium burns spontaneously in air. A mole of oxygen molecules is used up for every two moles of MgO formed. There are less arrangements for the solid product than the gaseous reactant, so we might expect the entropy to decrease, giving a negative value for ΔS.

It is quite easy to work out the exact value of ΔS for this reaction, using standard entropy values from the *Book of data*.

Standard entropy values:

$$S^{\ominus}[\text{Mg(s)}] = 32.7 \text{ J K}^{-1}\text{ mol}^{-1}$$
$$S^{\ominus}[\tfrac{1}{2}\text{ O}_2(\text{g})] = 102.5 \text{ J K}^{-1}\text{ mol}^{-1}$$
$$S^{\ominus}[\text{MgO(s)}] = 26.9 \text{ J K}^{-1}\text{ mol}^{-1}$$

(Notice how much higher the standard entropy of the gas, oxygen, is than that of the solids, magnesium and magnesium oxide.)

We can calculate $\Delta S^{\ominus}$ for the reaction in much the same way as we found $\Delta H^{\ominus}$ values using energy cycles earlier in this Topic.

$$2\text{Mg(s)} + \text{O}_2(\text{g}) \longrightarrow 2\text{MgO(s)}$$
$$\begin{aligned}
\Delta S^{\ominus} &= 2S^{\ominus}[\text{MgO(s)}] - 2S^{\ominus}[\text{Mg(s)}] - 2S^{\ominus}[\tfrac{1}{2}\text{O}_2(\text{g})]\\
&= 2 \times 26.9 - 2 \times 32.7 - 2 \times 102.5\\
&= -216.6 \text{ J K}^{-1}\text{ mol}^{-1}
\end{aligned}$$

$\Delta S^{\ominus}$ is indeed negative, showing that the entropy of the reaction system has decreased. Yet the reaction nevertheless goes spontaneously. Why?

2 *The reaction of hydrogen with oxygen*   When hydrogen and oxygen are mixed together and ignited, they react explosively. Since this reaction makes two molecules where there were three before, the entropy can be expected to decrease, giving a negative value of ΔS. Once again, the value of $\Delta S^{\ominus}$ is easily calculated from standard entropy values. It comes to -326.4 J K^{-1} mol^{-1}.

$$2\text{H}_2(\text{g}) + \text{O}_2(\text{g}) \longrightarrow 2\text{H}_2\text{O(l)}; \qquad \Delta S^{\ominus} = -326.4 \text{ J K}^{-1}\text{ mol}^{-1}$$

3 *The freezing of water*   When water freezes, a liquid turns to a solid. There are more arrangements for the relatively free molecules in liquid water than for the molecules in ice, so the entropy decreases.

$$H_2O(l) \longrightarrow H_2O(s); \qquad \Delta S^{\ominus} = -22 \text{ J K}^{-1} \text{ mol}^{-1}$$

Here, then, are three changes with negative ΔS values which nevertheless happen spontaneously. (For the last example, the freezing of water, whether or not it happened spontaneously would depend on where you lived and perhaps the time of year.) The negative ΔS values show that the number of ways of arranging molecules and their energies is getting *less*, but nevertheless the changes occur. Why?

The clue to the answer lies in the *enthalpy* values for each of the changes. In each case ΔH is negative – the changes are all exothermic. When the change occurs, energy is given out and passed to the *surroundings* – the air, container, or whatever. Now, this energy will have an effect on the entropy of these surroundings. Extra quanta of energy are being made available, and of course this increases the number of ways energy quanta can be arranged among the molecules in the surroundings. In other words, the exothermic change increases the entropy of the surroundings, and $\Delta S_{surroundings}$ is positive. As we shall see later, $\Delta S_{surroundings}$ can be calculated. When we consider the *total* entropy change of any process, we will always have to take into account $\Delta S_{surroundings}$ as well as ΔS_{system}, which is the entropy change of the chemicals themselves. In other words,

$$\Delta S_{total} = \Delta S_{system} + \Delta S_{surroundings}$$

For a process in which ΔS_{system} is negative, $\Delta S_{surroundings}$ may be sufficiently positive to cancel it out, giving an overall positive total entropy change. For example, for the burning of magnesium,

$$\Delta S_{system} \quad = -216.6 \text{ J K}^{-1} \text{ mol}^{-1}$$
$$\Delta S_{surroundings} = +4038 \text{ J K}^{-1} \text{ mol}^{-1} \text{ (at 298 K)}$$

Therefore,

$$\Delta S_{total} = -216.6 + 4038 \text{ J K}^{-1} \text{ mol}^{-1}$$
$$= +3821.4 \text{ J K}^{-1} \text{ mol}^{-1}$$

This large positive value ensures the reaction happens spontaneously.

Calculating the entropy change in the surroundings

Chemical reactions usually involve quite large enthalpy changes. Therefore in most reactions $\Delta S_{surroundings}$ is quite substantial and certainly cannot be ignored. But how can we calculate its value? We cannot possibly count all the ways, W, of sharing the energy among the surroundings – for one thing, the surroundings are impossible to define exactly.

Fortunately, there is a simple relation which enables us to calculate $\Delta S_{surroundings}$ easily. It is

$$\Delta S_{surroundings} = \frac{-\Delta H_{reaction}}{T}$$

where T is the temperature of the surroundings. The exact derivation of this relation is complicated and outside the scope of this book. But it is quite easy to get a feeling for why the relation has this form.

For a reaction whose enthalpy change is $\Delta H_{reaction}$, a quantity of energy $(-\Delta H_{reaction})$ is passed to the surroundings. The minus sign arises because, if the system loses the energy, the surroundings gain it, and vice versa. So you can see why we have the $(-\Delta H_{reaction})$ term: the more energy passed to the surroundings, the greater the increase in number of ways of sharing energy, and so the greater the entropy change.

But why divide by T? Well, if T is high and the surroundings are already hot, giving them some more energy will not make much difference to the entropy – there is already plenty of energy to share. But if T is low and the surroundings are cold, passing energy to them will make a bigger difference to the entropy, and will multiply the sharing possibilities a lot. In other words, the entropy change will vary *inversely* with temperature. That is why we divide by T.

In Topic 4 we saw a simple example. For two molecules:

Number of quanta shared	W, number of ways of arranging quanta	Adding 1 quantum multiples W by
1	2	—
2	3	1.50
3	4	1.33
4	5	1.25
10	11	1.10

So, if a system already has a lot of quanta of energy, the fractional increase in the number of ways when you add another one is less than if it started with little energy.

After all, someone on a low salary appreciates a £10 rise more than someone on a high one. But they would both prefer a £20 rise to a £10 one.

How to make reactions go the way you want

Let us now consider some applications of these ideas. We have already looked at the explosive reaction of hydrogen with oxygen.

$$2H_2(g) + O_2(g) \longrightarrow 2H_2O(l); \quad \Delta S^{\ominus} = -326.4 \text{ J K}^{-1} \text{ mol}^{-1}$$
$$\Delta H^{\ominus}_{298} = -571.6 \text{ kJ mol}^{-1}$$

From what we have said, we would expect $\Delta S_{\text{surroundings}}$ to be large and positive enough to overcome the negative ΔS_{system} of -326.4 J K^{-1}mol^{-1}. Using the formula

$$\Delta S_{\text{surroundings}} = \frac{-\Delta H}{T},$$

the calculation is quite simple, though we must remember to convert ΔH from kilojoules to joules. If the surroundings are at 298 K, then

$$\Delta S_{\text{surroundings}} = \frac{-(-571\ 600 \text{ J mol}^{-1})}{298 \text{ K}}$$
$$= 1918 \text{ J K}^{-1} \text{ mol}^{-1}$$

This is more than enough to make up for the negative ΔS_{system}. In fact

$$\Delta S_{\text{total}} = \Delta S_{\text{system}} + \Delta S_{\text{surroundings}}$$
$$= -326.4 + 1918 \text{ J K}^{-1}\text{mol}^{-1}$$
$$= +1591.6 \text{ J K}^{-1}\text{mol}^{-1}$$

Thus, when the surroundings are taken into account, the increase in entropy of the surroundings easily counterbalances the decrease in entropy of the reacting system.

We shall return to this kind of balancing act later, in Topics 10 and 12 in which some equilibrium processes are discussed.

For the moment, we just note that a reaction which is exothermic is likely to involve an overall increase in entropy. Having energy to give to the surroundings will certainly increase the number of ways of distributing the energy in the surroundings. Equally, a reaction which is endothermic has to borrow energy from the surroundings, reducing the number of ways of distributing the energy there. It will need some entropy on the credit side if it is to take place: maybe by splitting molecules into a larger number of fragments.

How can we help to obtain the products we want? The temperature of the surroundings is a factor, and one we can do something about.

What do you do if you want the 'reaction' of turning liquid water into ice cubes to happen? You put the water in the freezing compartment of the refrigerator, in cold surroundings. Changing water to ice involves a decrease of entropy of 22 J K^{-1} mol^{-1}, as fewer ways of sharing energy exist in the solid; $\Delta S_{\text{system}} = -22$ J K^{-1} mol^{-1}.

However, for $H_2O(l) \longrightarrow H_2O(s)$, $\Delta H^{\ominus}_{298} = -6$ kJ mol^{-1}.

Therefore 6000 J mol^{-1} are transferred to the surroundings during the process. Is this enough to make ΔS_{total} positive? Let's try it.

At 300 K $\Delta S_{\text{surroundings}} = \dfrac{-(-6000)}{300} = +20$ J K^{-1} mol^{-1}

This is not enough to counterbalance the negative entropy change of the system ($\Delta S_{\text{system}} = -22$ J K^{-1} mol^{-1}). The total entropy change is negative, and the water does not freeze.

At 250 K $\Delta S_{\text{surroundings}} = \dfrac{-(-6000)}{250} = +24$ J K^{-1} mol^{-1}

This is more than enough to counterbalance ΔS_{system}. The total entropy change is now positive, and the water freezes.

The *same* energy raises the entropy *more* if the surroundings are *cooler*. So to encourage an exothermic process, don't have things too hot. For example

$$2H(g) \longrightarrow H_2(g)$$

is very exothermic, and on the fairly cool surface of the Earth all the hydrogen we know is in the form H_2. Any H atoms we make soon combine. But on the Sun, there is a lot of H and essentially no H_2. The surroundings are too hot, the entropy change of the surroundings for the combining reaction is small, and other entropy changes win.

Equally, if you want to give an endothermic reaction a good chance, make the surroundings *hot*. Then the entropy decrease, as the reaction takes energy from the surroundings, will be less than if the surroundings were cooler.

For many chemical reactions, ΔH is big enough to make the entropy change in the surroundings more significant at normal temperatures than any entropy changes in the reaction system. That is why exothermic reactions are so much more common than endothermic ones. In the exothermic direction, the entropy change in the surroundings will always be positive, and this will usually be enough to counterbalance any negative entropy changes in the system. So we can talk of 'energetic stability' and of exothermicity as a guide to the direction of spontaneous change. Endothermic reactions can, and do, happen, but they *always* have a large, positive entropy change in the reaction system, to overcome the entropy decrease in the surroundings.

By carefully altering the reaction conditions, particularly temperature, we can beat the trend and make reactions go in the direction *we* want. The rules described below show how to do this.

To summarize the ideas from the work so far (Topics 3, 4, and 6).

1 If you are considering a particular reaction –

It helps to make the reaction happen if	*It does not help to make the reaction happen if*
the reaction produces more molecules than it started with, because that increases the number of ways	the reaction produces fewer molecules than it started with, because that reduces the number of ways
the reaction gives energy to the surroundings, since that increases the number of ways of sharing energy in the surroundings	the reaction takes energy from the surroundings, since that reduces the number of ways of sharing energy in the surroundings

2 To make a reaction go the way you want, you should

Raise the pressure if it produces fewer molecules than at the start;
lower the pressure if it produces more molecules than at the start.
Whatever the reaction, see if you can have a lot of reactant and few product molecules present.
If the reaction is exothermic, cool it (but beware, because that makes it hard to borrow energy to cross a barrier, and may make the reaction so slow that it barely happens).
If the reaction is endothermic, heat it.

BACKGROUND READING 2
Man and his energy sources

Man has always needed energy and, from the earliest times up to the present, much of it has been produced by the breaking and making of chemical bonds. Man by himself has only his own muscles, driven by the energy released when food molecules are oxidized, giving him a power of about 30 W. In the course of history he learned to release energy by the oxidation of fuels. He used wood first. Later coal, a concentrated energy source, enabled him to use fuel to produce mechanical as well as thermal energy. Coal powered the Industrial Revolution in Britain. The earliest steam engines, such as Newcomen's of 1712, had a power of about 4000 W (4 kW), and by 1900 steam engines had been developed to give a power output of 9000 kW. Man had enormously expanded the power of his arm, and in doing so had found the means to exploit the World's resources to raise his own standard of living.

Today a typical large modern fossil-fuelled power station would be rated at 2000 MW, but despite the huge change in magnitude of power output, we still largely depend, as did primitive Man, on the fact that carbon and hydrogen atoms form stronger bonds to oxygen than they do to each other. Most of our energy still comes from the oxidation of organic molecules present in coal, oil, and natural gas. Our appetite for energy is prodigious: the total annual consumption of energy in the United Kingdom is about 10^{19} J. This averages over the whole population to give every man, woman, and child in the country a power consumption of about 6 kW: equivalent to 6 electric fires burning day and night.

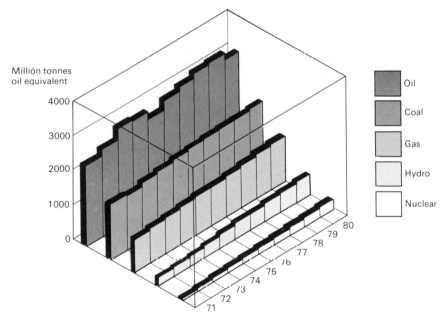

Figure 6.9
World consumption of primary energy, 1971 to 1980.
Based on The British Petroleum Co. Ltd BP statistical review of the World oil industry 1980, 1981.

Figure 6.9 shows how the World consumption of different energy sources changed in the period 1971–1980. The most noticeable feature of the graphs is the increase in consumption of oil in relation to solid fuel (mainly coal). Oil is more convenient than coal. It is easier to get out of the ground because it can be pumped, and it is easier to transport and distribute. Liquid fuels derived from petroleum are more convenient for powering cars, lorries, trains, and even power stations than solid fuels such as coal. In the sixties and early seventies, when oil was cheap and abundant, the developed world came more and more to depend on oil as its chief energy source. It was only in the seventies that most people began to realize that oil, and eventually the other fossil fuels, must at some time, run out.

	Coal /million tonnes coal equivalent	Oil /million tonnes
Australia	76	21
Canada	23	76
People's Republic of China	373	94
Federal Republic of Germany	120	5
India	72	12
Latin America	—	239
Middle East	—	1 114
Poland	167	—
Republic of South Africa	73	—
Soviet Union	510	546
United Kingdom	108	38
United States	560	467
Other regions	368	455
TOTAL WORLD	2 450	3 067

Table 6.2a
World production of coal and oil in 1977. (Thermal equivalent – 1.5 tonnes coal = 1 tonne oil.) *Note:* 'million tonnes coal equivalent' refers to all forms of hard coal (*i.e.*, excluding peat and lignite).

	Coal /million tonnes coal equivalent	Oil /million tonnes
Canada	4 242	900
People's Republic of China	98 883	2 800
Soviet Union	109 900	8 600
United Kingdom	45 000	2 300
United States	166 950	4 100
Other regions	237 957	70 200
TOTAL WORLD	662 932	88 900

Table 6.2b
World reserves of coal and oil, as estimated in 1980. *Tables 6.2a and b are from The British Petroleum Company Ltd (1981)* BP Statistical review of the World oil industry 1980 *and* World coal study *(1980) Ballinger Publishing Company, Cambridge, Mass.)*

Tables 6.2a and b give some idea of how much coal and oil is produced in different parts of the World and how much is left in reserve. Of course, production rates are bound to change and it is certain that further reserves, as yet unknown, will be discovered. (These, however, may prove more expensive to exploit than present ones.) Nevertheless, the figures demonstrate strikingly how little oil the World still has. Britain, with its North Sea offshore oil reserve, is in a relatively favourable position but it is estimated that this will not last long after the end of this century. The World is running out of oil and this is mainly due to consumption by the Western countries. Energy consumption in different parts of the globe is grossly uneven; the United States, with six per cent of the World's population, uses twenty-seven per cent of its energy.

The depletion of the World's oil supplies and the fact that the major consumers of oil are not the major producers have resulted in an energy crisis. It is not just that the developed countries have economies that depend largely on oil as an energy source. Oil is also our main source of the organic raw materials needed for the manufacture of plastics, dyes, drugs, paints, man-made fibres, detergents, printing inks, antifreeze, agricultural chemicals, and the countless other products of the petrochemical industry. The need to conserve remaining oil reserves is of overwhelming urgency for the future of the World.

However, oil cannot be conserved unless it is replaced by another source of energy. In the short term we have enough reserves of coal, provided that we can develop the technology to enable us to substitute it for oil. In the long term, however, we will have to realize that we have come to the end of an era of fossil fuels, and that we will no longer be able to rely on the energy obtainable from breaking and making chemical bonds. We will need to look for alternative sources of energy which are renewable, for by then our supplies of non-renewable chemical energy will have gone.

Look at table 6.3 and imagine yourself 300 years ahead; what energy sources will the World be using by then?

Renewable	Non-renewable
Solar source	Coal
Wind	Oil
Wave	Natural gas
Tide	Nuclear fission (uranium)
Geothermal source	
Biomass (wood, etc.)	
Solid wastes	
Nuclear fusion (deuterium)	

Table 6.3
Renewable and non-renewable sources of energy. Renewable sources are constantly being replenished or are so large that they will never be used up. Non-renewable sources are finite and irreplaceable.

BACKGROUND READING 3
Food

Foods contain many nutrients we all need for health. But only a few supply energy. These are fats, carbohydrates, and proteins. The first two of these are concerned directly with the supply of energy. Fats and, to a lesser extent, carbohydrates, can be stored in the body; when required they are converted into carbon dioxide and water, and thus their energy is released. Although a high proportion of the protein in foods has a structural function (it helps to make muscles, skin, bone, liver etc.), the cells of the structures are constantly being turned over. In other words, each day a certain proportion of the cells in the tissues is removed and broken down, and replaced by new cells. As cells are broken down their protein becomes available as a source of energy. Thus, in effect, all food proteins eventually become a supply of energy for the body.

The two most important factors which determine energy requirements are metabolic rate and level of physical activity. Even two apparently similar people can have very different energy needs.

The next table below shows the average amount of energy needed daily by different groups of people if they are to maintain ideal 'weight'.

Age /years	Occupational category	Energy /MJ	Age /years	Occupational category	Energy /MJ
Boys			*Men*		
1		5.0	18–34	sedentary	10.5
2		5.75		moderately active	12.0
3–4		6.5		very active	14.0
5–6		7.25	35–64	sedentary	10.0
7–8		8.25		moderately active	11.5
9–11		9.5		very active	14.0
12–14		11.0	65–74	sedentary	10.0
15–17		12.0	75 +	sedentary	9.0
Girls			*Women*		
1		4.5	18–54	most occupations	9.0
2		5.5		very active	10.5
3–4		6.25	55–74	sedentary	8.0
5–6		7.0	75 +	sedentary	7.0
7–8		8.0	Pregnant		10.0
9–11		8.5	Lactating		11.5
12–14		9.0			
15–17		9.0			

Table 6.4
From DHSS (1981) Recommended daily amounts of food energy and nutrients for groups of people in the United Kingdom.
Report on Health and Social subjects No. 15, H.M.S.O.

If a person consistently eats less energy-producing foods than he needs he loses body fat (and some protein) and loses 'weight'. Some people who eat more than they need metabolize the excess energy and remain slim. But many convert the excess energy in proteins, fats, and carbohydrates to body fat and store it. They become obese.

Although it is recommended that we should measure food energy in joules, calories are more familiar to many people.

An approximate conversion is 1 kilocalorie = 4.18 kJ.

The energy value of a food depends on

the mass eaten

the proportion of fats, proteins, and carbohydrates in that food

These energy-yielding nutrients provide different amounts of energy per unit mass.

1 g fats provides 37 kJ

1 g proteins provides 17 kJ

1 g carbohydrates provides 16 kJ

It is interesting to note that the alcohol in 1 g of 70° proof spirits, as in 1 g of gin or whisky, provides 29 kJ.

So if fat-rich foods make up a high proportion of the diet, this diet is likely to contain a large amount of energy. You can see the contribution different foods make to the total energy intake in the following example. Notice how little energy most vegetables provide – and how much butter, pastry, and other fatty foods contain.

	Energy	
	kJ	kcal
Breakfast		
20 g cornflakes (4 tablespoons)	313	74
150 cm^3 milk	408	98
10 g sugar (2 level teaspoons)	168	39
40 g toast (1 slice)	506	119
10 g butter	304	74
20 g marmalade	222	52
Mid-morning		
50 g chocolate biscuits (2)	1 099	262
Lunch		
50 g beefburger (1)	542	130
75 g bun (1)	743	175
15 g salad vegetables	9	2
150 g chips	1 598	380
200 cm^3 Coca-cola (1 can)	336	78

Energy

	kJ	kcal
Evening		
150 g steak and kidney pie	1 908	457
50 g peas (2 tablespoons)	111	26
50 g sliced beans (2 tablespoons)	15	3
200 g boiled potato (2 medium)	686	160
125 g apple crumble	1 098	260
50 g custard (3 tablespoons)	248	59
40 g sugar (8 teaspoons in 4 cups of tea)	672	156
200 cm^3 milk in tea/coffee	544	130
TOTAL	11 530	2 734

BACKGROUND READING 4
Rocketry

Rocket propellants consist of two classes of substances, *fuels* and *oxidizers*.

A fuel may be a single element, such as hydrogen, a compound such as hydrazine, N_2H_4, or a mixture of compounds such as kerosine. An oxidizer need not necessarily consist of, or even contain, oxygen, but can be any chemical element or compound having an electronegative character. Fluorine and the other halogens, sulphur, and to a lesser extent nitrogen and phosphorus can all be classed as oxidizers. In addition, a whole range of compounds containing these elements behaves as oxidizers; some examples are nitric acid, ammonium chlorate(VII), and potassium manganate(VII).

The chemical reactions taking place between fuels and oxidizers may be simple, as in the case of hydrogen and fluorine:

$$H_2(g) + F_2(g) \longrightarrow 2HF(g); \quad \Delta H = -542 \, kJ \, mol^{-1}$$

fuel   oxidizer

or fairly complex as in the UDMH/dinitrogen tetroxide reaction (UDMH stands for unsymmetrical dimethylhydrazine):

$$2(CH_3)_2N\!-\!NH_2(l) + 3N_2O_4(l) \longrightarrow 4CO(g) + 8H_2O(g) + 5N_2(g);$$

fuel   oxidizer

$$\Delta H = -2416 \, kJ \, mol^{-1}$$

In addition there are a few substances that can act as propellants by their own decomposition, without the need for an oxidizer. These are called *monopropellants*. An example is hydrogen peroxide, which decomposes to a hot mixture of steam and oxygen.

$$H_2O_2(l) \longrightarrow H_2O(g) + \tfrac{1}{2}O_2(g); \qquad \Delta H = -54.2\,\text{kJ}\,\text{mol}^{-1}$$

A number of factors determine the choice of propellants, and these include the density of the fuel and oxidizer, the properties of the exhaust gases, and the *specific thrust* (that is, the thrust per unit mass flow rate) obtainable from the reaction taking place. The last of these in turn depends upon several factors, including the temperature and the average molar mass of the exhaust gases and is sometimes called the *specific impulse*. We shall now consider these factors.

Density Propellants must be stored in tanks forming part of the rocket vehicle. If the propellant has a low density, large (and therefore heavy) tanks are required to contain a given quantity of propellant. It follows that the vehicle will be less efficient than one employing denser propellants contained in a smaller and lighter tank arrangement. To be considered at all as propellants, gases such as oxygen and hydrogen must be liquefied. The technical problems posed by the storage of these very cold liquids ($T_b = 90\,\text{K}$ and $20\,\text{K}$ respectively) are formidable. They can be stored only in a boiling condition, and continual tank topping is necessary.

Properties of the exhaust gases If a rocket is powered by hydrogen peroxide the reaction products are water and oxygen. Both of these are components of the atmosphere and do not therefore constitute a hazard. On the other hand, hydrogen and fluorine form an excellent fuel and oxidizer combination from many points of view, but the consequences of showering a rocket launching site with tonnes of hydrogen fluoride make this combination impossible to use!

Thrust In a rocket motor the specific thrust is produced by the stream of hot gas molecules that leave its nozzle. The rocket motor produces this thrust whether or not it is surrounded by air, and it is in fact the only known means of obtaining thrust in the vacuous conditions of outer space; its thrust is actually enhanced when it operates in the absence of a surrounding atmosphere. If a rocket motor is to produce a steady thrust there must be a continuous consumption of propellant to sustain the flow of gases out of the nozzle. Obviously a propellant must be chosen so as to give as high a thrust for as small a consumption of propellant as possible.

The specific thrust depends upon the exhaust velocity of the gases, and this can be calculated from the equation

$$v_e = \sqrt{\frac{2C_pT_c}{M}\left[1 - \left(\frac{T_e}{T_c}\right)\right]}$$

where

v_e is the exhaust velocity of the gases
C_p is the molar heat capacity of the exhaust gases
T_c is the combustion chamber temperature
M is the average molar mass of the exhaust gases
T_e is the temperature of the exhaust gases

Note that the average molar mass of the exhaust gases appears in this equation. If the specific thrust is to be as great as possible, the average molar mass of the exhaust gases must be as small as possible: another factor to consider when choosing a propellant.

There are about a dozen elements or compounds available which form suitable rocket oxidizers and several dozen which are possible fuels. This means there are between 500 and 1000 possible propellant combinations. To test each combination in order to establish its performance would be a formidable task, as regards both time and expense. It is better, if possible, to calculate the performance of all these combinations and then, perhaps, test by experiment a few of those that appear most promising. The next tables give some data.

LIQUID PROPELLANTS

Oxidizer	Fuel	Average density of propellant $/g\,cm^{-3}$	Combustion temperature T_c/K	Mean molar mass of exhaust gases	v_e $/m\,s^{-1}$
Oxygen (liquid)	Ethanol	0.99	3390	24.1	2739
	Hydrazine	1.07	3400	19.3	3079
	Hydrogen	0.28	3000	10.0	3841
	Kerosine	1.02	3670	23.3	2941
	UDMH	0.98	3580	21.3	3048
Fluorine (liquid)	Hydrazine	1.31	4680	19.4	3566
	Hydrogen	0.45	3870	11.8	4023
Hydrogen peroxide (95%)	Hydrazine	1.26	2856	19.5	2767
	Kerosine	1.30	3277	22.1	2678
	UDMH	1.24	2828	21.7	2727
Dinitrogen tetroxide	Kerosine	1.25	3450	25.7	2709
	UDMH	1.18	3410	23.6	2798

SOLID PROPELLANTS

Composition	Average density of propellant /g cm^{-3}	Combustion temperature T_c/K	Mean molar mass of exhaust gases	v_e /m s^{-1}
Black powder (potassium nitrate, carbon, and sulphur)	2.10	3500	98.0	1371
Cordite (cellulose nitrate + glyceryl trinitrate)	1.59	2388	26.0	2759
Composite (ammonium chlorate(VII) and organic polymer)	1.76	3013	26.0	2469
Composite and aluminium additive	1.80	3784	31.0	2605

Let us calculate the exhaust velocity obtained from a propellant which does not produce too high a combustion chamber temperature and which gives reaction products that are stable compounds or elements. This is the simplest class to deal with and is represented by the monopropellant hydrogen peroxide, which decomposes as follows:

$$H_2O_2(l) \longrightarrow H_2O(g) + \tfrac{1}{2}O_2(g)$$

First, we calculate the mean molar mass of exhaust gases:

$$M = \frac{(1 \times 18) + (\tfrac{1}{2} \times 32)}{1\tfrac{1}{2}} = 22.65$$

Now we must calculate the combustion chamber temperature T_c. Thermochemical tables, such as those in the *Book of data*, list the standard enthalpy changes of formation of molecules at some reference temperature, usually 298 K. Such tables tell us that at 298 K the standard enthalpy changes of formation, $\Delta H_{f,298}^{\ominus}$, for $H_2O_2(l)$, $H_2O(g)$, and $O_2(g)$ are -187.8, -241.8, and 0.0 kJ mol^{-1} respectively. Therefore, if we assume that the decomposition of hydrogen peroxide in the rocket takes place at 298 K, the enthalpy change for the reaction will be

$$\Delta H = \Delta H_f^{\ominus}[H_2O(g)] - \Delta H_f^{\ominus}[H_2O_2(l)] = -54.0 \, \text{kJ mol}^{-1}$$

This is the amount of energy which would be evolved if the reaction did actually take place at 298 K. In the combustion chamber, however, the 54.0 kJ of heat are retained by the products of the reaction, which means that their temperature will certainly be higher than 298 K. How can we find this temperature? We must seek the temperature at which the enthalpy of a mixture of $H_2O(g) + \frac{1}{2}O_2(g)$ is 54.0 kJ higher than at 298 K. Thermochemical tables are available which list the differences in enthalpy of substances between a temperature T K and 298 K, that is, $H_T^\ominus - H_{298}^\ominus$. Here are some values of $H_T^\ominus - H_{298}^\ominus$ for the substances in which we are interested.

Temperature /K	$(H_T^\ominus - H_{298}^\ominus)$/kJ mol^{-1} $H_2O(g) + \frac{1}{2}O_2(g)$
1000	37.38
1100	43.32
1200	49.44
1300	55.64
1400	62.10

This table shows, for example, that if the enthalpy of $H_2O(g) + \frac{1}{2}O_2(g)$ is 49.44 kJ above that at 298 K, the temperature of the system is 1200 K. Use the table to find the temperature in the combustion chamber of the rocket.

We now have values for T_c and M but before we can proceed to calculate v_e we still need values for C_p and T_e. The calculation of T_e involves more advanced thermodynamic principles, so let us cheat a little and tell you that $T_e = 502$ K and that $C_p = 41.74$ J mol^{-1} K^{-1}. Using the equation, you can now find the effective exhaust velocity of the rocket.

To help you with this calculation remember that, using SI units, 1 joule = 1 newton metre where the newton is that force which imparts an acceleration of 1 metre s^{-2} to a mass of 1 kilogram. If, therefore, in the equation, you substitute $C_p = 41.74 \times 10^3$ J kg^{-1} mol^{-1}, your answer will be in m s^{-1}. As a further exercise, calculate the combustion chamber temperature of a rocket using pure hydrazine as propellant, assuming the overall reaction may be represented as

$$N_2H_4(g) \longrightarrow \frac{1}{2}N_2(g) + \frac{1}{2}H_2(g) + NH_3(g)$$

The standard enthalpy changes of formation are as follows.

Substance	$\Delta H_{f,298}^\ominus$ /kJ mol^{-1}
$N_2H_4(g)$	+50.6
$NH_3(g)$	-46.1
$H_2(g)$	0
$N_2(g)$	0

The variation in product enthalpies with temperature is given in the following table.

Temperature /K	$(H_T^{\ominus} - H_{298}^{\ominus})/\text{kJ mol}^{-1}$		
	$NH_3(g)$	$N_2(g)$	$H_2(g)$
1300	50.16	31.55	29.95
1400	56.48	34.94	33.10
1500	63.01	38.46	36.37
1600	69.72	41.98	39.64

Answer: $T_c = 1463\,\text{K}$

High temperature calculations The example of the decomposition of hydrogen peroxide was deliberately chosen because it has a low combustion temperature and steam and oxygen as products. As a result of this low temperature the specific thrust produced by a given mass of propellant is correspondingly poor. The more energetic propellant combinations give rise to combustion temperatures several times as high as the hydrogen peroxide decomposition temperature. At these high temperatures the phenomenon known as chemical dissociation in the products occurs.

Chemical dissociation is little more than the failure of a reaction to go to completion, as it would if the products cooled to more moderate temperatures. Instead, a certain proportion of the products is dissociated into reaction intermediates, normally considered to have only very brief lifetimes. A typical example of chemical dissociation is the following process.

$$H_2(g) \rightleftharpoons H(g) + H(g); \qquad \Delta H = +436.0\,\text{kJ mol}^{-1}$$

The reaction is written with an equilibrium sign rather than a single arrow, because the dissociation is a dynamic equilibrium. At a given temperature, a portion of the molecular hydrogen will be dissociated into atomic hydrogen and, the higher the temperature or the lower the pressure, the greater will be the extent of this dissociation. The effects of pressure and temperature on an equilibrium system can be qualitatively predicted by Le Châtelier's principle. At first sight dissociation might appear beneficial, since its products possess a lower molar mass than their parent molecule.

However, dissociation involves the breaking of chemical bonds, which is an endothermic process. This causes a reduction in the flame temperature which more than offsets the reduction in molar mass. Here is a case where only one equilibrium has to be taken into account; in a rocket with hydrogen and oxygen as the propellant the reaction, allowing for all possibilities, is:

$$4H_2 + O_2 \longrightarrow aH_2O + bH_2 + cO_2 + dOH + eH + fO$$

where a, b, c, d, e, and f are the proportions of the mixture. To determine a, b, c, d, e, and f and to find the combustion temperature, all the possible equilibria must be considered and a set of simultaneous equations solved. We must also bear in mind that equilibrium constants vary with temperature, in a non-linear manner. The H_2/O_2 reaction is a simple one, yet its solution could well occupy a chemist for a whole day even if all the relevant charts and tables of thermochemical data were available. With a typical solid propellant consisting of ammonium chlorate(vii) as the oxidizer and a polyurethane resin as the fuel, no less than thirteen chemical species are found in the combustion chamber. Clearly, the human effort involved in solving the problems of the associated equilibria would be enormous and it is more than likely that fatigue would set in well before a solution was achieved!

Fortunately, by feeding computers with the thermochemical data, we can calculate accurately the performance of large numbers of fuel–oxidizer combinations.

The types of propellant which have been selected for each of the various rockets used in space exploration by the United States are shown in the table on page 188.

Figure 6.10
Apollo 17 Trans Lunar coast view of full Earth, from a distance of about 21 750 miles (35 000 kilometres) from Earth, December 1972.
Photograph, NASA/Space Frontiers Ltd.

Vehicle	No. of stages	Propellant	Approximate take-off thrust $/N \times 10^{-3}$
Scout	4	solid	440
Thor-Delta	3	lox/kerosine WFNA/UDMH solid	670
Thor-Agena	2	lox/kerosine RFNA/UDMH	730
Atlas D	$1\frac{1}{2}$	lox/kerosine	1 630
Atlas-Agena	$2\frac{1}{2}$	lox/kerosine RFNA/UDMH	1 630
Titan II	2	NTO/UDMH hydrazine	1 910
Atlas Centaur	$2\frac{1}{2}$	lox/kerosine lox/hydrogen	1 630
Saturn I	3	lox/kerosine lox/hydrogen lox/hydrogen	6 670
Saturn V	3	lox/kerosine lox/hydrogen lox/hydrogen	37 830
Space Shuttle	$2\frac{1}{2}$	lox/hydrogen solid lox/hydrogen	26 000

Table 6.6
NASA space launching vehicles
lox – liquid oxygen
WFNA – white fuming nitric acid
RFNA – red fuming nitric acid
UDMH – unsymmetrical dimethylhydrazine
NTO – dinitrogen tetroxide

SUMMARY

At the end of this Topic you should:

1 know the meanings of the terms
   exothermic reaction and endothermic reaction
   standard enthalpy change of a reaction
   standard enthalpy change of formation of a compound
   standard enthalpy change of atomization of an element
   standard enthalpy change of combustion of an element or compound;
2 be able to measure the enthalpy changes of some reactions, using either
   an electrical compensation calorimeter or some other piece of apparatus;
3 know Hess's Law, and be able to apply it to find enthalpy changes that
   cannot be measured directly;
4 be able to calculate the standard enthalpy change of a reaction from the
   standard enthalpy changes of formation of the reactants and products;
5 be able to use the food and fuel calorimeter;
6 know the meaning of the term bond energy, and be able to use values
   of bond energies;
7 be aware of the use of the bomb calorimeter in accurate thermochemistry;
8 understand the Born–Haber cycle, and be able to use it to do calculations;
9 know what is meant by lattice energy, and be able to relate theoretical
   and experimental values of lattice energies;
10 know, and be able to use, the relationship between entropy change
   and enthalpy change,

$$\Delta S = -\Delta H/T$$

11 be aware of World energy sources and their likely period of life, and
   have some idea of World energy requirements;
12 be aware of the applications of energy studies, particularly in food and
   in rocket propulsion.

PROBLEMS

*Indicates that the *Book of data* is needed.

1 $100 \, cm^3$ of 0.02M copper(II) sulphate solution were put in an electrical
  compensation calorimeter and an excess of magnesium powder was
  added. 1052 joules had to be supplied to give the same rise in temperature
  as that which resulted from the reaction.

a Calculate the moles of copper(II) ions used.
b Calculate $\Delta H^\ominus$ for the reaction in $kJ \, mol^{-1}$.

c Is heat evolved or absorbed during the reaction?
d Write the equation for the reaction, together with the enthalpy change, giving it the correct sign.

2 50 cm^3 of 0.05M silver nitrate solution were placed in an electrical compensation calorimeter and an excess of copper powder was added. 184 joules had to be supplied to give the same rise in temperature as that which resulted from the reaction.

a Calculate the moles of silver ions used.
b Calculate $\Delta H^{\ominus}$ for the reaction in kJ mol^{-1} of copper used.
c Write the equation for the reaction, together with the enthalpy change, giving it the correct sign.
d For what length of time would a current of 0.1 ampere at a potential of 12 volts need to be passed in order to give the same temperature rise as that which resulted from the reaction?

3 Using your results from questions 1 and 2, calculate the enthalpy change per mole of magnesium atoms when magnesium reacts with silver nitrate solution. Write the equation for the reaction, together with the enthalpy change, giving it the correct sign. What assumptions have you made in your calculation?

*4 Ethanol burns in an excess of air according to the equation

$$C_2H_5OH(l) + 3O_2(g) \longrightarrow 2CO_2(g) + 3H_2O(l).$$

Calculate the value for the enthalpy change for this reaction in the following way.

a Write down the equation for the reaction.
b Draw a diagram showing the formation of the compounds on both sides of the equation from the same elements, as shown on page 154.
c Using the *Book of data*, find the standard enthalpy changes of formation for the products, and write them in the correct place on the diagram.
d Do the same for the reactants.
e Calculate the required enthalpy change, giving it the correct sign.

5 1 g of each of the following alcohols was burned in a calorimeter similar to that shown in figure 6.5. In each case the quantity of energy required to give the same temperature rise as in the reaction was determined; the energy was supplied electrically.

Alcohol	Energy required
Methanol CH_3OH	22.34 kJ
Ethanol C_2H_5OH	29.80 kJ
Propanol C_3H_7OH	33.50 kJ
Butanol C_4H_9OH	36.12 kJ

In each case calculate the standard enthalpy change of combustion of the alcohol.

Comment on the results you obtain and, using graphical means, make a prediction of the enthalpy change of combustion for pentanol, $C_5H_{11}OH$.

*6 Calculate the enthalpy changes at 298 K when the following changes occur. State whether the heat is given to or taken from the surroundings.

a 8.0 g of iron are added to an excess of a solution of a copper(II) salt. (Fe^{2+}(aq) ions are formed.)

b 24 dm^3 (298 K, 101 kPa) of an equimolar mixture of hydrogen and carbon monoxide are burned in oxygen. Assume that the water produced is liquid.

c 2.3 g of gaseous sodium atoms are converted to gaseous ions Na^+(g).

*7 Using the standard enthalpy changes of formation of the compounds, and those of atomization of the elements, calculate the enthalpy changes for the following reactions.

a $NH_3(g) \longrightarrow N(g) + 3H(g)$
b $PH_3(g) \longrightarrow P(g) + 3H(g)$
c $AsH_3(g) \longrightarrow As(g) + 3H(g)$
d $SbH_3(g) \longrightarrow Sb(g) + 3H(g)$

What generalizations do your answers indicate about the energies of the bonds in the hydrides of Group V elements?

*8 Using the standard enthalpy changes of formation of the compounds, and those of atomization of the elements, calculate the enthalpy changes for the following reactions.

a $HF(g) \longrightarrow H(g) + F(g)$
b $HCl(g) \longrightarrow H(g) + Cl(g)$
c $HBr(g) \longrightarrow H(g) + Br(g)$
d $HI(g) \longrightarrow H(g) + I(g)$

What generalizations do your answers indicate about the energies of the bonds in the hydrides of the Group VII elements?

*9 Draw a fully labelled Born–Haber cycle for the formation of calcium oxide. From this, calculate the lattice energy for calcium oxide.

*10 Use your *Book of data* to determine the lattice energy of sodium monoxide, Na_2O.
The oxide NaO does not exist. Explain why, and state what further thermodynamic data you would need to substantiate your argument.

*11 Calculate $\Delta H^{\ominus}_{298}$ for the following reactions.

a $CH_3OH(l) + 1\frac{1}{2}O_2(g) \longrightarrow CO_2(g) + 2H_2O(g)$
b $SO_2(g) + 2H_2S(g) \longrightarrow 3S(s) + 2H_2O(l)$
c $Ba^{2+}(aq) + SO_4^{2-}(aq) \longrightarrow BaSO_4(s)$
d $N_2O(g) + Cu(s) \longrightarrow CuO(s) + N_2(g)$
e $NH_4Cl(s) \longrightarrow NH_3(g) + HCl(g)$
f $NH_4Cl(s) \longrightarrow NH_4^+(aq) + Cl^-(aq)$
g $Mg(s) + \frac{1}{2}O_2(g) \longrightarrow MgO(s)$
h $Mg^{2+}(g) + O^{2-}(g) \longrightarrow MgO(s)$
i $CH_4(g) + 2O_2(g) \longrightarrow CO_2(g) + 2H_2O(l)$
j $CO_2(g) + 2Mg(s) \longrightarrow 2MgO(s) + C(s)$
k $Ag^+(aq) + Cl^-(aq) \longrightarrow AgCl(s)$
l $AgCl(s) \longrightarrow Ag^+(g) + Cl^-(g)$
m $Na(s) + \frac{1}{2}Cl_2(g) \longrightarrow NaCl(s)$
n $NaCl(s) \longrightarrow Na^+(aq) + Cl^-(aq)$
o $NaCl(s) \longrightarrow Na^+(g) + Cl^-(g)$
p $CH_4(g) \longrightarrow C(g) + 4H(g)$

12 Suppose you were given the enthalpy changes for the following reactions.

$2Fe(s) + 1\frac{1}{2}O_2(g) \longrightarrow Fe_2O_3(s)$
$Ca(s) + \frac{1}{2}O_2(g) \longrightarrow CaO(s)$

What further information, if any, would you require in order to calculate the enthalpy changes of each of the following reactions?

a $3Ca(s) + Fe_2O_3(s) \longrightarrow 3CaO(s) + 2Fe(s)$
b $Ca(s) + CuO(s) \longrightarrow CaO(s) + Cu(s)$
c $2Fe(s) + 3CuO(s) \longrightarrow Fe_2O_3(s) + 3Cu(s)$

13 The enthalpy change of formation of sodium chloride over the temperature range 98–808 °C is about $-414\,kJ\,mol^{-1}$ whereas the enthalpy change of formation over the temperature range 808–892 °C is about $-385\,kJ\,mol^{-1}$. Explain why the enthalpy change of formation of sodium chloride changes abruptly at 808 °C by about $29\,kJ\,mol^{-1}$.

14 Given that:

$$P(g) + 3Cl(g) \longrightarrow PCl_3(g); \qquad \Delta H = -983 \, kJ$$
$$P(s) + 1\tfrac{1}{2}Cl_2(g) \longrightarrow PCl_3(g); \qquad \Delta H = -305 \, kJ$$
$$P(g) + 3H(g) \longrightarrow PH_3(g); \qquad \Delta H = -958 \, kJ$$
$$P(s) + 1\tfrac{1}{2}H_2(g) \longrightarrow PH_3(g); \qquad \Delta H = -8.4 \, kJ$$
$$P(s) \longrightarrow P(g); \qquad \Delta H = 314 \, kJ$$

calculate the bond energies of the following bonds.

a P—Cl in PCl_3
b P—H in PH_3
c Cl—Cl in Cl_2
d H—H in H_2

15 Ethanol and methoxymethane have the same molecular formula, C_2H_6O. The standard enthalpy change of combustion at 298 K of ethanol(g) is $-1367 \, kJ \, mol^{-1}$ and that for methoxymethane(g) is $-1460 \, kJ \, mol^{-1}$.

a Write equations for the combustion of (*i*) ethanol and (*ii*) methoxymethane, including the enthalpy changes for these reactions.
b Deduce $\Delta H^{\ominus}_{298}$ in $kJ \, mol^{-1}$ for the hypothetical change:

methoxymethane(g) $\longrightarrow$ ethanol(g)

c Suppose ΔH_1 is the enthalpy change for the change:

$$C_2H_6O \text{ (ethanol)(g)} \longrightarrow 2C(g) + 6H(g) + O(g)$$

and ΔH_2 is the enthalpy change for the change:

$$C_2H_6O \text{ (methoxymethane)(g)} \longrightarrow 2C(g) + 6H(g) + O(g)$$

Calculate $\Delta H_1 - \Delta H_2$.
d Suggest a reason for the difference in value between ΔH_1 and ΔH_2.

16 Consider the reaction $C(s) + O_2(g) \longrightarrow CO_2(g)$
a Use the *Book of data* to find the standard molar entropies of $C(s)$, $O_2(g)$, and $CO_2(g)$. Use this data to calculate the standard molar entropy change, $\Delta S^{\ominus}_{\text{reaction}}$, for the reaction.
b Use the *Book of data* to find the standard enthalpy change, $\Delta H^{\ominus}_{298}$, for the reaction. Use your answer to calculate the entropy change in the surroundings, $\Delta S^{\ominus}_{\text{surroundings}}$, at 298 K for the reaction.
c Calculate the total entropy change, $\Delta S^{\ominus}_{\text{total}}$, for the reaction.
d Does your answer to **c** suggest this reaction will be spontaneous at 298 K?

TOPIC 7
Structure

The chemical and physical properties of materials are strongly influenced by their structure at an atomic level. Therefore to understand the properties of materials it is necessary to understand their structures. This applies as much to naturally occurring substances such as rocks and minerals, and the constituents of living organisms such as cells, muscles, and bone, as it does to man-made substances such as alloys for high-speed turbine blades and polymers for drip-dry textile fibres. Indeed, before a new synthetic material can be designed it is necessary to have some knowledge of how molecular structure affects properties: then by synthesizing an appropriate structure it is possible to produce a substance with predetermined properties – a 'tailor-made' material.

7.1
SOME PHYSICAL METHODS FOR DETERMINING STRUCTURE

A very wide range of methods is available for obtaining information about the structures of substances; almost any physical phenomenon which is affected by a material can be made to yield evidence of the structure of that material. The principal categories of phenomena are those in which:

1 Electromagnetic radiation (figure 7.1) or a stream of particles such as electrons interacts with matter to give diffraction patterns (X-ray diffraction, electron diffraction).

2 Electromagnetic radiation is emitted or absorbed by matter, giving rise to emission or absorption spectra (infra-red absorption, nuclear magnetic resonance);

3 Matter interacts with an electric or a magnetic field (measurement of dipole moments).

Each of these phenomena gives different information about a substance, and when the evidence from several of them is added together it is often possible to obtain a detailed knowledge of its structure.

The diffraction of X-rays by electrons in atoms and molecules

As an example of the interaction of electromagnetic radiation with matter to give diffraction patterns, we shall consider X-ray diffraction, for this is the most precise and versatile method available for the determination of structure in solids.

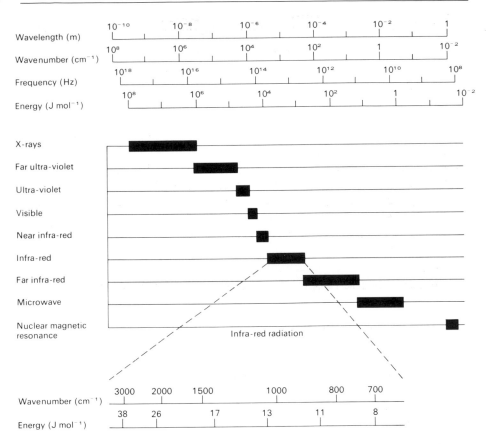

Figure 7.1
The relationship between wavenumber, wavelength, and energy of radiation.

X-rays have a wavelength in the region of 10^{-9} m, and are to be found in the electromagnetic spectrum beyond the far ultra-violet. When a beam of X-rays of one particular wavelength, that is, a monochromatic beam, falls on a crystalline solid the X-rays are scattered in an orderly manner. This scattering is known as *diffraction* and gives rise to a *diffraction pattern*, which can be recorded photographically or, with modern instruments, electronically. A typical X-ray diffraction pattern is shown in figure 7.2.

The X-ray diffraction pattern is related to the pattern of the electrons in the solid.

electromagnetic radiation whose frequency is the same as any of the vibrations in the molecule. The vibrations in the molecule will increase in amplitude as a result of the energy absorbed from the radiation. However, there is an important restriction to this behaviour: the absorption of radiation only occurs if the vibration is accompanied by a change of dipole in the molecule. A dipole is formed by two electric charges of equal magnitude but opposite sign, a small distance apart. Thus, molecules such as hydrogen and chlorine will not absorb radiation as a result of their molecular vibration, but polar molecules such as hydrogen chloride will absorb radiation.

The radiation that is absorbed as a result of molecular vibrations lies in the infra-red region of the electromagnetic spectrum. Particular vibrations in particular bonds give rise to absorption in a particular part of this region.

In recordings of infra-red spectra (figure 7.5) these absorptions are seen as deep troughs (inverted peaks).

When describing infra-red spectra the position of peaks is referred, not to the corresponding wavelength but to the reciprocal of the wavelength, known as the *wavenumber*, measured in cm^{-1}.

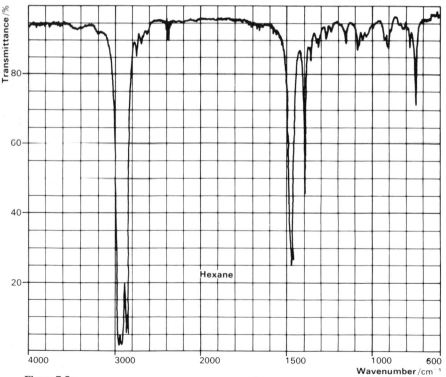

Figure 7.5a
The infra-red spectrum of hexane (thin film).

Individual wavenumbers in an infra-red spectrum are useful because each peak is characteristic of a vibration of a particular molecular structure. Thus the C—H stretching vibration absorbs at about $2900\,cm^{-1}$ in alkanes but at about $3050\,cm^{-1}$ in alkenes. We can see this in the infra-red spectra of hexane, $CH_3CH_2CH_2CH_2CH_2CH_3$ (figure 7.5a) and cyclohexene ◯ (figure 7.5b).

C⚌C vibrations in the plane of a benzene ring absorb at both $1600\,cm^{-1}$ and $1500\,cm^{-1}$, and this can be seen in the infra-red spectrum of poly(phenylethene), figure 7.6, overleaf.

The characteristic absorptions that are useful for the identification of particular groups of atoms in molecules of organic carbon compounds are found in the region from $200\,cm^{-1}$ to $4000\,cm^{-1}$. The vibrational wavenumbers of different bonds in organic molecules will be considered in the appropriate organic chemistry Topic 9, 11, or 13, and summarized in the final organic chemistry Topic 17.

Infra-red absorption spectra are obtained by using an infra-red spectrometer. A description of this instrument is given in the Background reading at the end of this section.

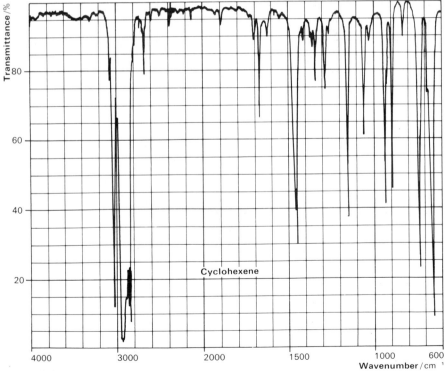

Figure 7.5b
The infra-red spectrum of cyclohexene (thin film).

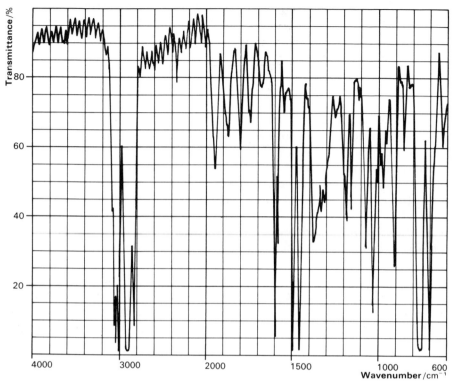

Figure 7.6
The infra-red spectrum of poly(phenylethene).

Nuclear magnetic resonance spectroscopy

Neutrons and protons, just like electrons, can be considered to possess spin. Whether a particular nucleus has a spin or not depends on the exact arrangement of its constituent neutrons and protons. All nuclei are electrically charged, and so for those with spin, the rotation of the charge about the spin axis produces an electric current and hence an associated magnetic field. Such nuclei can be considered to behave like small magnets. Typical examples are ^{1}H, ^{13}C, ^{19}F, and ^{31}P.

If these 'magnetic nuclei' are placed in an external magnetic field, they line up with this field (position A in figure 7.7). If energy is now supplied to system A, there is another allowed orientation for the nuclei, that is, with the nuclear magnet opposed to the external field (position B in figure 7.7). The energy difference ΔE between these two positions is very small and gives rise

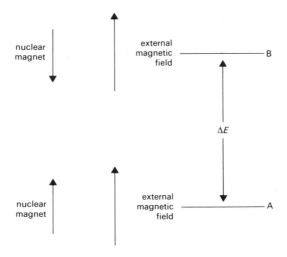

Figure 7.7

to absorption in the radio region of the electromagnetic spectrum.

This effect can be investigated experimentally by placing a sample of a compound containing 'magnetic nuclei' (for instance, CH_3OH or any other hydrogen compound) in a strong magnetic field and then irradiating the sample with radio waves (figure 7.8).

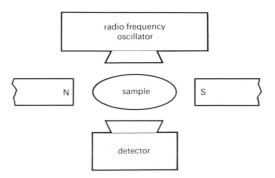

Figure 7.8

This is called *nuclear magnetic resonance spectroscopy*, NMR, and is like any other form of absorption spectroscopy. Electromagnetic radiation (in this case in the radio region) is absorbed only when the frequency corresponds exactly to ΔE. The value of ΔE, and hence the absorption frequency, depend not only on the applied magnetic field but also on the chemical environment of the nucleus. In this way, the nucleus is being used to 'probe' details of its chemical environment.

NMR is so sensitive that nuclei in slightly different chemical environments give rise to different absorption frequencies. For a complex compound, the spectrum of absorbed frequencies can be related back to different chemical environments of the nuclei.

In practice, it is easier to keep the radio frequency constant and change the applied magnetic field.

At the present time, NMR spectroscopy is one of the most valuable methods available to chemists for the determination of the structure of carbon compounds.

The effect of an electric field on molecules

We can carry out an experiment to see if molecules are affected by an electrostatic field.

EXPERIMENT 7.1
What is the effect of an electrostatic field on a jet of liquid?

Fill a burette with water, stand it over an empty beaker, and turn on the tap so that a stream of water flows into the beaker. Hold a charged glass rod near the jet of water, but do not let it touch the water. A glass rod can be charged by rubbing it vigorously with a piece of cloth, provided both are thoroughly dry.

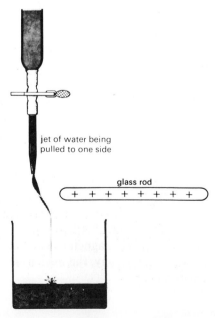

jet of water being pulled to one side

glass rod

+ + + + + + + +

Figure 7.9
Experiment to investigate the effect of an electrostatic field on liquids.

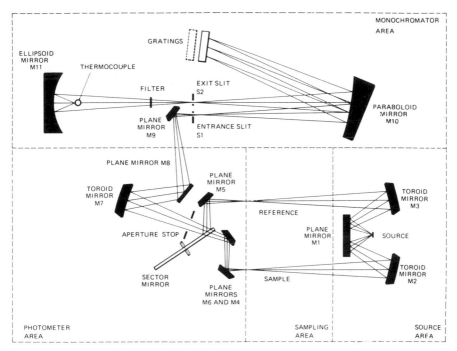

Figure 7.11
Layout of a typical spectrophotometer.
Diagram, Perkin-Elmer Ltd.

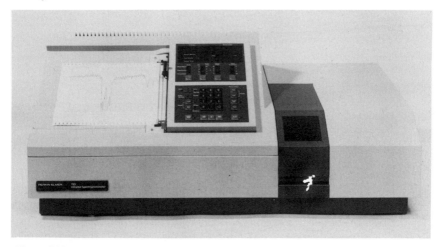

Figure 7.12
A laboratory spectrophotometer.
Photograph, Perkin-Elmer Ltd.

7.2
CRYSTAL STRUCTURES, AND CRYSTAL PROPERTIES

In this section we shall examine some models of crystal structure, and carry out some experiments to relate the properties of substances to their crystal structure. Some of the simplest crystal structures are those of metals: hexagonal close packing (h.c.p.); face-centred cubic packing (f.c.c.); and body-centred cubic packing (b.c.c.). Illustrations of these are shown in figure 7.13, 7.14, and 7.15.

In figure 7.13 the second layer is made by placing spheres in the hollows of the bottom layer. The third layer is made by placing spheres directly over the spheres in the bottom layer and a repeating pattern ABA is established. The fourth layer is formed by repeating the B arrangement. Thus a repeating pattern ABAB ... results; this is known as *hexagonal close packing*.

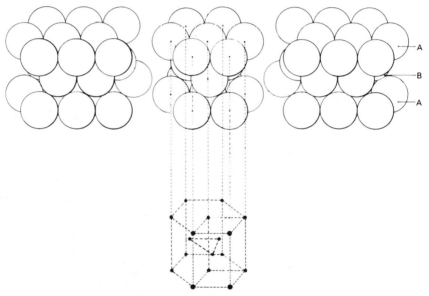

Figure 7.13
The hexagonal close-packed structure derived from layers of close-packed spheres in ABAB sequence, and its unit cell.

If one considers three adjacent spheres forming a triangle in the first B layer it is possible to arrange a similar triangle in the third layer in such a way that the spheres in the third layer do not lie directly above the bottom A layer (as they do in the h.c.p. structure). This is shown in figure 7.14. The third layer is thus a different arrangement, and could be called C. If the fourth layer is arranged to be a repeat of the bottom A layer, then the whole pattern may be repeated giving ABCABC...; this is known as *face-centred cubic packing*.

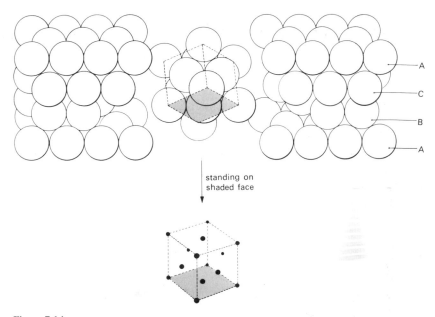

Figure 7.14
The face-centred cubic structure derived from layers of close-packed spheres in ABCA sequence, and its unit cell.

If you examine models of these structures that can be taken apart, you will find that in a hexagonal close-packed arrangement and in a face-centred cubic arrangement a given sphere will have 12 other spheres in contact with it. These are the closest possible packing arrangements for equal-sized spheres. For a given sphere there are 12 nearest neighbours, and it is said to have 12 co-ordination.

Figure 7.15 shows *body-centred cubic packing*: in this, each sphere is in contact with 8 others and is said to have 8 co-ordination. Thus, b.c.c. is not such a close-packed arrangement as h.c.p. and f.c.c. In h.c.p. and f.c.c. structures there is only 26 per cent of empty space but in the b.c.c. arrangement there is 32 per cent of empty space.

Almost all of the metals crystallize into one or more of these systems. Some examples are:

Hexagonal close-packed – magnesium, zinc, nickel
Face-centred cubic – copper, silver, gold, aluminium
Body-centred cubic – the alkali metals

The close-packed systems account for about fifty metals, and the body-centred system for about twenty metals. There is no obvious relation between the structural type of the metal and its position in the Periodic Table.

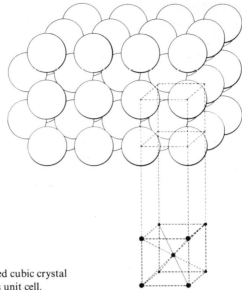

Figure 7.15
The body-centred cubic crystal
structure and its unit cell.

Now examine ball-and-spoke models of the crystal structures of some compounds. A number of these may be available to you, but they should certainly include sodium chloride and calcium fluoride. In each case work out the co-ordination number (number of nearest neighbours) for each ion.

Notice that in sodium chloride the chloride ions form a face-centred cubic arrangement. The sodium ions occupy what are known as *octahedral sites* between them. Let us now see what is meant by this term.

Look once again at the model of the face-centred cubic arrangement shown in figure 7.14. Pick any four atoms in one layer that touch one another. The centres of these four atoms, together with the centres of the two atoms that touch these four (one in the layer above, and one in the layer below) form the corners of a regular octahedron. At the centre of this octahedron is a hole, known as an *octahedral site*. Now suppose that these atoms are all ions having the same charge. An oppositely charged ion can be placed in this hole. This new ion is said to occupy an octahedral site, and will touch *six* of the first ions. Around any of the first ions there are *six* such octahedral sites. If you study the model of the sodium chloride structure you will see that sodium ions occupy all the available octahedral sites between the chloride ions. Each sodium ion has six chloride ions as its nearest neighbours (6 co-ordination) and each chloride ion is surrounded by six sodium ions (6 co-ordination).

Other possible sites exist, as can be seen from the model of the calcium fluoride structure. In this structure, notice that the calcium ions form a face-centred cubic arrangement. The other ions, in this case fluoride ions, occupy *tetrahedral sites* between the calcium ions.

To understand tetrahedral sites look again at the face-centred cubic arrangement of figure 7.14, and this time pick any three atoms in one layer that touch one another. The centres of these three atoms, together with the centre of the one atom in the next layer which touches all three, form the corners of a regular tetrahedron. At the centre of this tetrahedron is a hole known as a *tetrahedral site*. Now suppose that these atoms are all ions having the same charge. An oppositely charged ion can be placed in this hole. This new ion is then said to occupy a tetrahedral site, and will touch *four* of the first ions. Around any of the first ions there are *eight* such tetrahedral sites. If you study the model of the calcium fluoride structure you will see that fluoride ions occupy all the available tetrahedral sites between the calcium ions.

Octahedral and tetrahedral sites are to be found in both types of close-packed arrangements, f.c.c. and h.c.p.

A unit cell

Crystallographic studies are aided by considering a crystal to be made up of many adjacent identical 'unit cells'. A unit cell of sodium chloride is shown in figure 7.16. It consists of three ions in each edge of the cube. By convention each edge is chosen to contain two sodium ions and one chloride ion, but this is only a convention; two chloride ions and one sodium ion could equally well be chosen.

○ = sodium

⬀ = chlorine

Figure 7.16
A unit cell of sodium chloride.

Does the unit cell have an empirical formula NaCl? One may find an answer to this question by considering the extent to which the various ions in the unit cell are shared.

An ion at a *corner* is shared by 8 cells, giving $\frac{1}{8}$ ion per cell.
An ion on an *edge* is shared by 4 cells, giving $\frac{1}{4}$ ion per cell.
An ion on a *face* is shared by 2 cells, giving $\frac{1}{2}$ ion per cell.
An ion *inside* the cell is not shared, giving 1 ion per cell.

Thus in a unit cell of sodium chloride we have:

	Na$^+$	Cl$^-$
At the *corners*, 8 ions, $\frac{1}{8}$ charge each	1	
On the *edges*, 12 ions, $\frac{1}{4}$ charge each		3
On the *faces*, 6 ions, $\frac{1}{2}$ charge each	3	
Inside the cell, 1 ion		1
	$\overline{4}$	$\overline{4}$

The unit cell therefore contains the equivalent of four sodium ions and four chloride ions, giving an empirical formula of NaCl.

If you examine the unit cells of the two cubic arrangements of metals mentioned above, face-centred cubic (f.c.c.) and body-centred cubic (b.c.c.), you will see that in the close-packed arrangement eight atoms are situated at the corners of a cube, and six others are situated one at the centre of each of the six faces of the cube. The unit contains the equivalent of four atoms $(8 \times \frac{1}{8} + 6 \times \frac{1}{2})$. The *body-centred cubic structure* has eight atoms at the corners of the cube, one atom at the centre of the cube, but no others on the faces. This unit cell contains the equivalent of two atoms.

These two unit cells are shown in figure 7.17, and in figures 7.14 and 7.15.

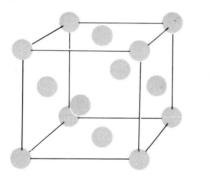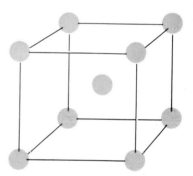

Figure 7.17
Face-centred and body-centred unit cells.

The Avogadro constant from X-ray evidence

X-ray measurements on crystals can be used to obtain a value for the Avogadro constant, L. For this, the dimensions of the unit cell, and the number of particles that it contains, are required, and each of these quantities can be determined accurately for many crystals. The number of particles which occupy the molar volume of the substance can then be calculated, as in the next example.

An end view of the unit cell of sodium chloride is shown in figure 7.18. From X-ray diffraction evidence the width of this unit cell is 0.5641 nm.

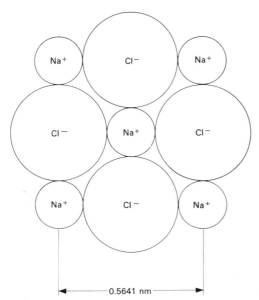

Figure 7.18
The unit cell of sodium chloride. This end view is of a space-filling type of model,
as distinct from the ball-and-spoke type of model shown in figure 7.16.

The unit cube contains 4NaCl, as explained on the opposite page.

The mass of one mole of sodium chloride, NaCl, is 58.44 g mol^{-1}
The density of sodium chloride is 2.165 g cm^{-3}

Therefore, the total volume of one mole of NaCl, including both ions and empty space, is

$$\frac{58.44 \text{ g mol}^{-1}}{2.165 \text{ g cm}^{-3}}$$

This contains L ion pairs of Na^+Cl^-. But the total volume share of one ion pair, that is, the volume of the two ions plus their share of empty space, is

$$\frac{(0.5641 \times 10^{-7})^3}{4} \text{ cm}^3$$

$$L = \frac{58.44 \text{ g mol}^{-1}}{2.165 \text{ g cm}^{-3}} \times \frac{4}{(0.5641 \times 10^{-7})^3} \text{ cm}^3$$

$$= 6.01 \times 10^{23} \text{ mol}^{-1}$$

This method is one of the most accurate for finding L, and has enabled its determination to be carried out to an accuracy of ± 0.01 per cent. The value of L is $6.022045 \times 10^{23} \, \text{mol}^{-1}$.

Physical properties and crystalline structures

In the following experiments we shall investigate the relationship between the physical properties of some crystals and their structures.

EXPERIMENT 7.2a
The cleavage of graphite and calcite

1 Use a magnifying glass to examine a small quantity of graphite powder. Note the flat, plate-like nature of the crystals.

Rub a little graphite powder between two fingers, and note the feel. The flat crystals slide readily over one another, markedly reducing the friction between the fingers. Graphite either on its own or suspended in oil is used as a lubricant.

Relate the appearance of the crystals to a model of the crystal structure.

2 Take a piece of calcite, and examine its shape and the cracks in it. What do you notice about the direction of the cracks?

Tap the crystal gently with a small hammer or the back of a closed penknife, or some similar instrument. What do you notice about the shape of the fragments?

Relate the shape of the fragments, and the directions of the cleavage cracks, to a model of the crystal structure.

EXPERIMENT 7.2b
Investigating the behaviour of some substances between crossed polaroids

Light waves from an electric lamp vibrate in an infinite number of different planes, at right angles to the line of propagation. A piece of polaroid has the property of transmitting only waves vibrating in one plane.

You are supplied with two pieces of polaroid mounted at right angles to each other, on a wooden bar. Support the polaroid assembly in a clamp stand so that the polaroid sheets are horizontal, and leave sufficient room below the sheets to place a torch bulb and holder there.

1 Place a lighted bulb beneath part of the assembly where there is only *one* piece of polaroid when viewed from above, that is, where the two sheets do not overlap. Now look at the bulb and polaroid through a loose piece of polaroid. Turn the loose piece round and round while you watch the bulb. Interpret

what you observe.

2 Examine a range of different materials as follows. Put the bulb under the centre of the crossed polaroids. Put a small sample (about $2 \times 2 \times 0.5$ mm) on a microscope slide. Insert the specimen between the crossed polaroids, until it is over the bulb. Rotate the specimen slowly, watching it carefully through the upper polaroid, and note whether any effect is produced or not.

Examine the following materials: sodium chloride, calcium carbonate (use a clear, transparent fragment of calcite), sodium iodide, potassium iodide, potassium thiocyanate, calcium fluoride (fluorite), quartz, and a 1–2 cm length of human hair.

Potassium thiocyanate, K^+NCS^-, is a substance whose structure you have probably not yet met. This structure is shown in figure 7.19.

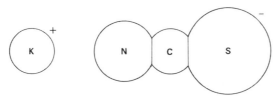

Figure 7.19
The structure of potassium thiocyanate.

Draw four columns in your notebook as shown below, and record your observations:

1	2	3	4
Name of substance giving no effect, or almost no effect, whatever the orientation	Structure of substance listed in column 1	Name of substance giving an effect which depends upon orientation	Structure of substance listed in column 3

From your knowledge of the shapes of the particles concerned, fill in columns 2 and 4. What do you notice? What conclusions can you draw about the types of structures which produce no effect, and the types which do produce an effect?

What do you think might be the shape of the molecular units which make up human hair?

Isotropy and anisotropy

An isotropic substance is one whose properties are the same in whatever direction they are measured. For instance, the refractive index of sodium chloride is the same whether light is passed through a crystal from side to side, top to bottom, or back to front; also the effect of sodium chloride upon polarized light is the same whatever the orientation of the crystal.

An anisotropic substance is one whose properties depend upon the direction in which they are measured. For instance, calcite has refractive indices varying from 1.49 to 1.66 depending upon the direction along which the light traverses the crystal. The thermal conductivity of calcite is also different in different crystallographic directions.

Isotropy can occur when the units which comprise a crystal are spherically symmetrical. Sodium ions and chloride ions are spherically symmetrical, and light waves or thermal vibrations are not affected differently because they approach a given ion from different directions. The carbonate ion is not spherically symmetrical, and both light waves and thermal vibrations experience different effects on approaching a given carbonate ion from different directions.

All crystals which adopt the cubic system are isotropic with respect to polarized light, the velocity of light in them (refractive index), thermal conductivity, and electrical conductivity; they are not isotropic to some phenomena, for instance the velocity of sound, and mechanical stretching (elasticity).

All substances which crystallize in other crystal systems are anisotropic.

These statements are consequences of the relation between spherical symmetry of the packing units and isotropy, for if a packing unit departs from spherical symmetry it cannot pack in a cubic system; the distortion produces some other system.

The two extremes of departure from spherical symmetry are planar units, and linear, rod-like units. The former are exemplified by the carbonate and nitrate ions, and by giant structures consisting of layers such as graphite, cadmium iodide, iron(III) choride, and the micas. The linear units are exemplified by the thiocyanate ion (NCS^-), by alkanes having unbranched carbon chains, and by fibrous structures such as asbestos, cotton, and hair.

The anisotropy of graphite

Look at a model of the structure of graphite. Is it symmetric or asymmetric? Would you expect graphite to be isotropic or anisotropic?

Because of its anisotropic properties much research has been conducted in attempts to produce large single pieces of graphite. One of the methods developed is a hydrocarbon gas-cracking process conducted above 2000 °C. It slowly produces a material called pyrographite, a dense metallic-looking substance in which all the graphite crystallites are similarly oriented.

Some of the physical properties of pyrographite are listed in the following table.

Property	Along the crystal planes	At right angles to the crystal planes
Tensile strength/N m^{-2}	124×10^6	34.5×10^6
Thermal conductivity/W cm^{-1} K^{-1}	2.0	0.025
Electrical resistivity/Ω cm	2×10^{-4}	2500×10^{-4}
Thermal expansion/K^{-1}	0.66×10^{-6}	20×10^{-6}

From these figures it will be seen that pyrographite is a highly anisotropic substance. The thermal conductivity along the crystal planes is almost 100 times greater than at right angles to the planes. The electrical conductivity is more than 1000 times greater along the planes than at right angles to them.

The highly anisotropic thermal conductivity has led to the use of pyrographite in rocket nozzles and for re-entry nose-cones. A layer of pyrographite enables heat to be conducted rapidly away from a zone at extremely high temperatures to low temperature areas; at the same time its poor conducting qualities at right angles provide protection to the rocket, or re-entry capsule.

7.3
STRUCTURE AND THE PERIODIC TABLE

You will recall the trends in structure across the Periodic Table which were noticed in section 1.2, during study of the physical properties of the elements across a period of the Periodic Table.

For convenience, some of the information is summarized below.

Element	Na	Mg	Al	Si	P(white)	S	Cl	Ar
Structural type	←— giant lattices —→				←——— molecules ———→			
Melting point/°C	98	650	660	1400	44	110	−100	−190
Enthalpy change of fusion/kJ mol^{-1}	2.6	8.9	11	46	0.63	1.4	3.2	1.2
Boiling point/°C	880	1100	2500	2400	280	440	−35	−190
Enthalpy change of vaporization/kJ mol^{-1}	89	130	290	380	12	9.6	10	6.5

|—— metals ——| |————— non-metals —————|

| metallic bonding | |——— covalent bonding ———|

Table 7.1
Some properties of the elements sodium to argon. (Numerical values are correct to two significant figures.)

With melting points and enthalpy changes of fusion there is a sharp break between silicon and phosphorus, reflecting a change from giant lattices to molecules.

An almost identical trend is seen with the boiling points and enthalpy changes of vaporization, and again the break and contrast are striking.

However, the arrangement of the atoms of the elements in hexagonal close-packed, face-centred cubic, or body-centred cubic structures does not depend in any regular way on the position of the elements in the Periodic Table.

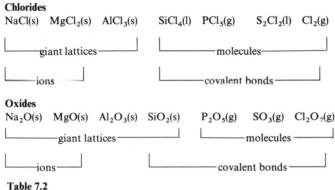

Chlorides
NaCl(s)   MgCl$_2$(s)   AlCl$_3$(s)   SiCl$_4$(l)   PCl$_5$(g)   S$_2$Cl$_2$(l)   Cl$_2$(g)

└───giant lattices───┘   └──────molecules──────┘

└──ions──┘   └────covalent bonds────┘

Oxides
Na$_2$O(s)   MgO(s)   Al$_2$O$_3$(s)   SiO$_2$(s)   P$_2$O$_5$(g)   SO$_3$(g)   Cl$_2$O$_7$(g)

└─────giant lattices─────┘   └────molecules────┘

└──ions──┘   └───covalent bonds───┘

Table 7.2
Structures of the chlorides and oxides of the elements sodium to argon.

In these compounds there is a transition from ionic bonds on the lefthand side of the table, through a type of bond intermediate between ionic and covalent bonds, to covalent bonds on the righthand side of the table.

This transition raises a number of questions. What are the natures of an ionic bond and of a covalent bond? How are the particles held together? Why does the transition occur? Answers to these questions may be found in terms of the electronic structures of the atoms, and in terms of electron rearrangements, and these will be discussed in the next Topic.

BACKGROUND READING 2
Minerals and their structure

The Earth's crust is a mixture of 3000 or more different minerals, many of which are 'impure' crystalline compounds. A study of crystal structure is therefore relevant in mineralogy and geology. Minerals in the form of ores and other economic mineral deposits are also important primary raw materials whose properties (like those of all solids) depend ultimately upon the spatial arrangements and bonding of their constituent atoms and ions. Such a study is thus further linked with the engineering disciplines – mining, mineral processing, metallurgy, and others. A brief account is now given of the impact of structure on some mineral properties of technological interest.

What is a mineral? Opinions differ, but most authorities would agree that a mineral is a naturally occurring crystalline solid having a characteristic range of properties which are constant or nearly so. This implies that minerals have an essentially definite composition and crystal structure, and, indeed, there is little difference except in purity between well known minerals such as rock salt (halite), fluorite (fluorspar), haematite, native copper, and limestone and the corresponding chemical compounds sodium chloride, calcium fluoride, iron(III) oxide, copper metal, and calcium carbonate to be found on the laboratory shelf. Of course, the majority of laboratory chemicals do not occur naturally, but general structural similarities are evident. A few mineral phases such as petroleum, air, water, coal, and volcanic glass cannot be considered, even approximately, as having a definite crystal structure under ordinary conditions. The term *mineraloid* is sometimes applied to these. Certain minerals, such as clays and feldspars, can exist over a considerable range of composition in so-called *solid solutions*.

A glance at any text of mineralogy or geology indicates the great variety of the natural inorganic compounds. They represent a multitude of shapes, colours, hardness, textures, and reactivities. The rock-forming minerals such as feldspars, quartz, and mica are abundant, while others like cinnabar (mercury sulphide) and native gold are scarce. They range from beautifully formed crystals of museum quality to certain more anonymous constituents in common soil.

Mineral groups

How can order be brought to so large a subject? One approach is to group minerals according to crystal structure. It is found from X-ray diffraction measurements, for instance, that rock salt (NaCl), galena (PbS), periclase (MgO), pyrite (FeS_2), and numerous other minerals all adopt the same cubic structure as sodium chloride: a face-centred cubic array of anions with all of the octahedral sites occupied by cations (tetrahedral sites being vacant). In each case the cation is seen to have about the right size to fit at the centre of six touching anions but is too large for a tetrahedral arrangement of four anions. The S_2^{2-} ion in pyrite functions as a single unit. Sphalerite (ZnS), greenockite (CdS), cinnabar (HgS), and chalcopyrite ($CuFeS_2$) have a similar cubic structure but with half of the tetrahedral sites occupied by cations (octahedral sites being vacant). Chalcopyrite can be written $Cu_{0.5}Fe_{0.5}S$, with each cation occupying half of the available positions in the cubic array. Table 7.3 gives a selection of ways in which minerals of widely differing general reactivity may be inter-related simply and elegantly on the basis of structure adopted.

Mineral	Basic structure	Octahedral	Tetrahedral	Stoicheiometry/ formula
Rock salt	f.c.c. (Cl^-)	All (Na^+)	Nil	$NaCl$
Sphalerite	f.c.c. (S^{2-})	Nil	Half (Zn^{2+})	ZnS
Fluorite	f.c.c. (Ca^{2+})	Nil	All $(2F^-)$	CaF_2
Braunite	f.c.c. $(2Mn^{2+})$	Nil	Three-quarters $(3O^{2-})$	Mn_2O_3
Perovskite	f.c.c. $(Ca^{2+} + 3O^{2-})$	One quarter (Ti^{4+})	Nil	$CaTiO_3$
Spinel	f.c.c. $(4O^{2-})$	Half $(2Al^{3+})$	One-eighth (Mg^{2+})	Al_2MgO_4
Wurtzite	h.c.p. (S^{2-})	Nil	Half (Zn^{2+})	ZnS
Corundum	h.c.p. $(3O^{2-})$	Two-thirds $(2Al^{3+})$	Nil	Al_2O_3
Ilmenite	h.c.p. $(3O^{2-})$	Two-thirds $(Fe^{2+} + Ti^{4+})$	Nil	$FeTiO_3$
Olivine	h.c.p. (O^{2-})	Half $(2(Mg^{2+}, Fe^{2+}))$	One-eighth (Si^{4+})	$(Fe,Mg)_2SiO_4$

Table 7.3
Occupation of interstitial sites in some minerals.

The huge silicate group of minerals – accounting for over 93 volume percentage of the Earth's crust – may also be represented in a simple ordered manner (types A–F, table 7.4), and some striking correlations then become evident between molecular constitution and the external properties of the minerals. An example of the simplest type of silicate is olivine. This is composed of tetrahedral anions (SiO_4^{4-}) in combination with cations (Mg^2 and Fe^{2+}). Note that the Si^{4+} ion is very small and highly polarizing and will therefore enter into predominantly covalent bonding with O^{2-} ions. Note also that the larger Mg^{2+} and Fe^{2+} ions are about the same size as each other and are interchangeable in any proportion in the olivine structure. In fact, they form a solid solution. The tetrahedral anions are able to condense together to give more complex anions, a process facilitated by the large bond energy $(466\,kJ\,mol^{-1})$ of the Si—O bond. This condensation is shown in the table in terms both of theoretical equations and tetrahedra sharing increasing proportions of their corners.

Omitting type B (of which few examples are known), the resulting structures take the form of discrete tetrahedra (A), chains of tetrahedra (C), double chains (D), sheets (E), and 3-D networks (F). Types A and F generally have strong covalent and/or ionic bonding throughout and are thus mechanically hard and difficult to cleave. Quartz consists solely of a symmetrical 3-D network of Si—O bonds. The mineral cannot be scratched with a knife. In contrast, pyroxenes and amphiboles (C and D) have relatively strong bonds in one direction only – along the length of the chains – and they have a fibrous structure in consequence. An example is tremolite asbestos, whose double chains cleave readily

Table 7.4
Some silicate structures.

Type	Example	Mode of formation and structure
A: discrete tetrahedra	$(Fe,Mg)_2SiO_4$ olivine	$SiO_4^{4-} \equiv \left[\begin{array}{c} O \\ O \cdots Si \cdots O \\ O \end{array} \right]^{4-}$
B: dimeric tetrahedra	Ca_2MgSiO_7 melilite	$2SiO_4^{4-} + 2H^+ \xrightarrow{-H_2O} Si_2O_7^{6-}$ one corner shared
C: chains of tetrahedra (pyroxenes)	$MgSiO_3$ enstatite	$nSiO_4^{4-} + 2nH^+ \xrightarrow{-nH_2O} (SiO_3^{2-})_n$ two corners shared
D: double chains (amphiboles)	$Ca_2Mg_{95}(Si_4O_{11})_2(OH)_2$ tremolite	$2(SiO_3^{2-})_n + nH^+ \xrightarrow{-0.5n\,H_2O} (Si_4O_{11}^{6-})_{0.5n}$ two and three corners shared
E: sheets of tetrahedra (micas and clays)	$Al_4Si_4O_{10}(OH)_8$ kaolinite	$n(SiO_3^{2-})_n + n^2H^+ \xrightarrow{-0.5n^2H_2O} (Si_4O_{10}^{4-})_{0.25n^2}$ three corners shared
F: stacks of sheets (feldspars and quartz)	SiO_2 quartz	$n(Si_4O_{10}^{4-})_{0.25n^2} + n^3H^+ \xrightarrow{-0.5n^3H_2O} (SiO_2)_{n^3}$ four corners shared

to give long fibres. Type E minerals – clays and micas – have weak bonds in one direction only. Kaolinite, for instance, consists of uncharged sheets held together by residual (van der Waals) bonding. This material has an earthy texture and is easily broken up between the fingers.

Crystal growth and dissolution

Imagine the process of formation of a mineral under aqueous or molten conditions of crystallization. We have a dynamic situation in which two main events occur together: solvated cationic and anionic species bombard the crystal surface and some lose their solvation shell to become part of the lattice (in such proportion as to maintain electrical neutrality); and some lattice ions shake loose, are solvated, and move away into the liquid medium. So long as the concentration of species in solution is great enough to ensure that the first of these events occurs more rapidly than the second, the mineral crystal grows (or *vice versa* dissolves). Eventually, at equilibrium, the rates become equal. As a particular type of crystal surface (plane) has uniform properties peculiar to itself, it will grow evenly and at a different rate from neighbouring planes. Thus, crystal surfaces tend to be flat and to grow relatively rapidly in surface area if their rate of (outward) growth is slower than that of neighbouring planes. This is an example of anisotropic behaviour.

Now, X-ray diffraction measurements suggest that the unit cell of rock salt is a cube of side 0.56 nm. Notwithstanding the above discussion, it is helpful to think about macroscopic salt crystals as a very large number of such cells stacked like building bricks so as to reproduce the observed external shape or *habit* of the mineral. Macroscopic crystals might then be cubic, as in common salt recrystallized from water, but could have any shape depending upon the prevailing 'building plan'. Remember that the Pyramids were built from rectangular slabs of rock! Recrystallization of rock salt from urea solutions gives octahedra. In this case the organic molecules are assumed to adsorb preferentially on the triangular octahedral faces, causing them to enlarge in surface area at the expense of the square cubic faces.

Defects

Of course, in practice crystallization processes do not yield perfectly formed crystals, even when they occur under closely controlled conditions. At best in rock salt about one in every million sites is vacant at room temperature, and there are likely to be numerous *dislocations* – slight misplacements of planes or blocks of ions on the nanometre scale – in the structure, which can cause mechanical weakening by a factor of one hundred or more and lead to a general increase in reactivity.

Natural conditions give rise to a whole host of additional defects. The most elementary studies of geography and geology suggest that particular minerals and mineral aggregates (rocks) can differ greatly from one location to another, not only in general abundance, but also in a whole range of chemical and physical properties, despite having basically the same crystal structure and bonding. Wide variations in habit occur because external physical constraints in a reaction cavity interfere with the 'ideal' growth of crystal planes. The results are often particles of all shapes and sizes containing massive cracks together with dissolved or physically trapped impurities. As natural crystallization generally takes place from a complex mother liquor, several minerals may precipitate simultaneously and give rise to an intimately intergrown aggregate. This may become fractured during cooling or subsequent geological processes to yield a complex mixture of composite particles. Even minerals of apparently simple composition, *e.g.* SiO_2, can vary in appearance from colourless and transparent to black and opaque, and in structure from symmetrical and well crystallized to amorphous and glassy. Thus we have the different named forms: quartz, chalcedony, agate, flint, opal, etc. Evidently, ideal properties governed by crystal structure and bonding are greatly influenced in practice by both the composition and the physical condition of the crystallizing liquor. It is the particular skill of the mineralogist or geologist to recognize any residue of ideal character in order to make positive identifications of minerals in the laboratory and field.

For these reasons, the 'hard sphere' model of ions packing with minimum energy and volume does not provide a ready basis for the *quantitative* explanation of mineral properties. We cannot, for example, use it to calculate precisely, from values of lattice or bond energies, the relative hardnesses of two minerals with the same face-centred cubic structure, such as pyrite and rock salt. It is not even possible to predict with certainty that rock salt should adopt a face-centred rather than a body-centred structure at room temperature – the two forms are separated in energy by a mere $10 \, kJ \, mol^{-1}$. On the other hand, crystal structures provide a good framework for investigating the various defects, in particular those occurring at the surfaces of crystals (where there must necessarily be discontinuities). In the case of diamond, for instance, which has a three-dimensional cubic lattice rather like that of sphalerite, it is anticipated that at the surface (to a depth of a nanometre or so) it is likely to resemble the hexagonal layering observed in graphite. The mineral is also likely to undergo oxidation in air to form surface hydroxyl, carbonyl, or epoxy groups. Indeed, there is thermochemical and electron diffraction evidence that a variety of organic reactions can take place on the surface of diamond, although these changes are not normally visible to the eye.

Processing

The bulk separation and purification of saleable minerals and metals are the concern of the mining and mineral processing industry and are based on practical mineral science. In particular, operations depend crucially upon the efficient identification and exploitation of *differences* in properties between minerals mixed together in natural deposits. Physical, chemical, and surface chemical properties are all used, so that economic separations can be made on the basis of hardness, shape, density, solubility, or surface 'wetability', to mention but a few examples. These technological considerations are taken up together with geochemical analogues in the Special Study *Mineral process chemistry*.

SUMMARY

At the end of this Topic you should:

1 know something of the use of the following methods for the determination of molecular and crystal structure –

X-ray diffraction, infra-red absorption spectroscopy, nuclear magnetic resonance spectroscopy, and measurement of dipole moments;

2 be able to recognize b.c.c., h.c.p., and f.c.c. structures;

3 know the structures of sodium chloride and calcium fluoride;

4 understand what is meant by tetrahedral and octahedral sites;

5 understand what is meant by a unit cell, and be able to work out the empirical formula from the structure of a simple type of unit cell;

6 be able to calculate a value for the Avogadro constant, given the dimensions of a simple unit cell;

7 be able to relate ease of cleavage, and rotation of the plane of plane-polarized light, with crystal structure;

8 know the meanings of the terms isotropy and anisotropy, and be able to give examples of both;

9 know the relationship between structural type of element, oxide, and chloride and position of element in the Periodic Table;

10 be aware of the importance of structure in the classification of minerals and the understanding of their properties.

PROBLEMS

1 The following are data relating to crystalline potassium bromide, K^+Br^-, which has a structure similar to that of sodium chloride.

Internuclear distance K^+ to Br^-	0.3285 nm
Density	2.75 g cm^{-3}
Mass of 1 mole	119.01 g

a Select a convenient crystal unit cell and *either* make a space-filling model of it *or* draw a sketch of it.

b What are the dimensions of the unit cell?

c How many ion pairs does the unit cell contain?

d Calculate the total volume share of one ion pair.

e Using the density and mass of 1 mole of K^+Br^- calculate the total volume (ions plus empty space) of one mole of K^+Br^-.

f Using the total volume share of one ion pair (from part **d**) and the total volume share of L ion pairs (from part **e**), calculate a value for the Avogadro constant.

g In calculating the Avogadro constant, which of the following limits most severely the accuracy of your estimate?
The accuracy of

A the K^+ to Br^- internuclear distance

B the density of potassium bromide

C your method of computation (state whether by calculator, slide rule, logarithm tables, etc.)

D both A and C

E both B and C

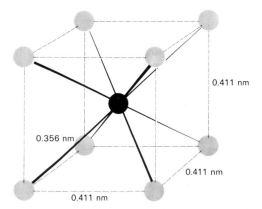

Figure 7.20
A unit cell of caesium chloride.

2 Figure 7.20 shows a unit cell of caesium chloride with a caesium ion at its centre.

a Which of the following most precisely describes this type of crystal structure?

A giant lattice

B double-face-centred cubic

C body-centred cubic

D hexagonal close-packed

E double simple cubic

b Using only the information supplied, deduce the empirical formula of caesium chloride. Is your answer consistent with the positions of caesium and chlorine in the Periodic Table?

c What is the co-ordination number of **(i)** the caesium ions and **(ii)** the chloride ions?

d The ratio of the radii of caesium and chloride ions is 0.93. That of sodium and chlorine in sodium chloride is 0.52. In the light of this, is the difference in co-ordination number of the ions in caesium chloride from that in sodium chloride to be expected or not? Explain how you arrive at your answer.

e Given the following additional data on caesium chloride, calculate a value for the Avogadro constant.

Density $3.988 \, \text{g cm}^{-3}$

Mass of 1 mole $168.36 \, \text{g mol}^{-1}$

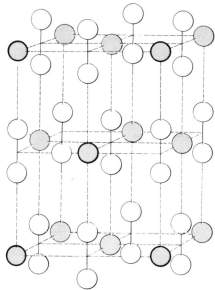

Figure 7.21
The crystal structure of calcium carbide.

3 Figure 7.21 shows the crystal structure of calcium carbide, $Ca^{2+}C_2^{2-}$.

a Which of the following structures does this crystal *most closely* resemble?

A graphite type

B simple cubic

C body-centred cubic

D hexagonal close-packed

E sodium chloride type

Would you expect this crystal to show isotropy or anisotropy? Explain why you answer as you do.

b

Compound	$\Delta H_{f,298}$/kJ mol^{-1}
$CaC_2(s)$	-60
$Ca(OH)_2(aq)$	-1004
$H_2O(l)$	-286
$C_2H_2(g)$	$+228$

Use these data to calculate the standard enthalpy change for the reaction between calcium carbide and water, producing ethyne gas and calcium hydroxide. Is the reaction exothermic or endothermic?

c Anhydrous barium peroxide $Ba^{2+}O_2^{2-}$ has a similar crystal structure to that of $Ca^{2+}C_2^{2-}$ and its reaction with water is exactly parallel also. Using this information, describe its reaction with water by means of an equation.

Bonding

In Topic 7 we studied the structure of materials, that is, the relative positions in space of the atoms in the molecules, or crystal lattices, of the substances under investigation.

In such studies one is at once led to consider also what holds the structures together. In other words, what is the nature of the interatomic and interionic forces?

Any theory of bonding which is to be of use must provide at least a reasonably satisfactory means of accounting for the formulae of compounds, for their structure, and for the nature of the forces which hold them together. Many of the properties of materials can be interpreted by simple models of how the electrons are distributed in the compounds. On the other hand there are properties, such as the electrical conductivity of metals, which can only be interpreted by more sophisticated models; and there are some properties for which there is as yet no adequate explanation.

In this Topic we shall examine simple models of electron distribution in compounds, and the ways in which they can account for the formulae and structure of compounds, and the forces which hold them together.

8.1
ELECTRON ARRANGEMENTS IN IONS

In your notebook, write down the evidence which you have encountered for the existence of ions. Can you also remember the names of any compounds that you have made that contain ions?

Typical compounds containing ions are lithium chloride, lithium oxide, magnesium fluoride, and magnesium oxide.

The ions are held together in crystals by the forces of electrostatic attraction between oppositely charged (+ and −) ions. The lattice energy (see Topic 6) of the crystalline compound is a measure of the strength of this attractive force.

The diagrams in figure 8.1 show how the transfer of electrons from one atom to another gives ions. In these diagrams the nucleus of each atom is represented by its symbol, and the shells of electrons are represented by groups of dots and crosses around the nucleus. The shell of lowest energy is nearest to the nucleus, and successively higher energy levels are shown at increasing distances. Although dots and crosses are used for the electrons in different atoms, it should not be thought that these electrons are distinguishable; this is merely

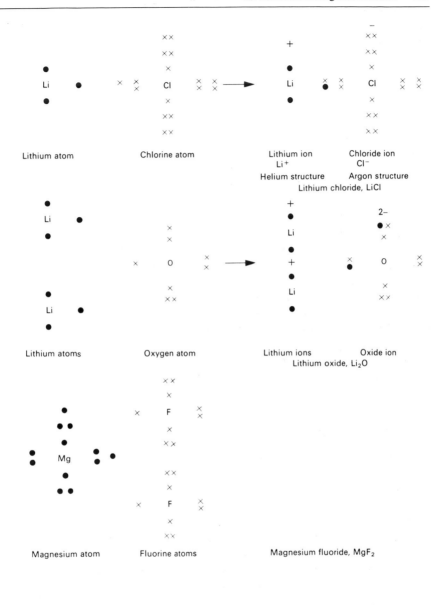

Figure 8.1
The formation of ions.

a device used in the diagrams to enable their movements to be followed.

It will be seen from the first two diagrams that the electronic structures of the ions that are formed are identical to those of a noble gas. These diagrams also show the formation of lithium chloride and lithium oxide. Study them, and then copy the last two, unfinished, diagrams into your notebook. Complete them, showing the electron transfers for these other two compounds, and write in the names of the noble gases whose structures are formed in these two cases.

You may like to draw similar diagrams to show how the following compounds might be formed by electron transfer: potassium oxide, and lithium hydride.

You will notice that making up the noble gas structure leads to the experimentally determined empirical formula.

It should be remembered when writing 'dot-and-cross' diagrams that the dots and crosses are a means of counting electrons, and showing the number present; they do not show the positions of the electrons. The electrons are distributed in space as diffuse negative charge-clouds.

It should also be noted that not all ions have a noble gas structure. The ions formed by d-block elements are cases in point. The Cu^{2+} ion, for example, has an electronic configuration $(1s^2)(2s^2 2p^6)(3s^2 3p^6 3d^9)$.

What elements form ions?

In order to form positive ions the element must be ionized; that is, electrons must be removed from its atoms. This requires energy. For a resulting compound to be stable the lattice energy must be large enough to compensate for the energy involved in ionization. This is commonly achieved for M^+ and M^{2+} compounds, but not for M^{3+} or M^{4+}. For example, the first and second ionization energies of magnesium are 740 and $1500 \, kJ \, mol^{-1}$ respectively, and magnesium will form ionic compounds. But the successive ionization energies of carbon are 1090, 2400, 4600, and $6200 \, kJ \, mol^{-1}$, and the very large amount of energy involved in ionizing carbon cannot be recovered in any lattice energy. Consequently carbon does not form C^{4+} ions; it does not in fact form stable positive ions at all but forms bonds by another method.

In order to form negative ions, atoms must gain electrons. The enthalpy change which takes place when gaseous atoms of an element acquire electrons and form single negatively charged ions is known as the *electron affinity* of that element. Some values of electron affinity are as follows.

Equation	Electron affinity /kJ mol^{-1}
$Cl(g) + e^- \longrightarrow Cl^-(g)$	-349
$Br(g) + e^- \longrightarrow Br^-(g)$	-325
$I(g) + e^- \longrightarrow I^-(g)$	-295
$O(g) + e^- \longrightarrow O^-(g)$	-141
$O^-(g) + e^- \longrightarrow O^{2-}(g)$	$+798$

From these figures you will see that the formulation of the various halide ions from atoms is exothermic, but the formation of oxide ($O^{2-}(g)$) ions from oxygen atoms ($-141 + 798 = 657\,\text{kJ mol}^{-1}$) is endothermic. The formation of $X^{3-}(g)$ ions, where X is any element, is strongly endothermic.

Can some ions be regarded as spheres?

When building models of ionic structures, those ions which are formed from single atoms are usually represented as spheres. It might be argued that as ions possess a complete outer electron shell it would be reasonable to suppose that the electron distribution is spherical. There is some experimental evidence for this view, based on a study of electron density maps, obtained by X-ray diffraction measurements. Figure 8.2 shows such electron density maps for sodium chloride and calcium fluoride.

Do the maps suggest that these ions are discrete entities, and if so, are the ions spherical?

How large are ions?

From figure 8.2 try to find the radius of a sodium ion and of a chloride ion. What difficulty is involved? Suggest one method of overcoming it. What do you think limits the accuracy of ionic radii found from electron density maps? What distance does a map of this nature give accurately?

In order to compile a table of ionic radii, one ionic radius has to be fixed arbitrarily; then other radii must be obtained. In addition to this problem, the size of the ion of an element varies slightly, depending upon the compound it is part of; ions are slightly soft, compressible, and deformable. For these reasons, tables of ionic radii compiled by different authorities do not always agree with each other. In spite of these uncertainties, the concept of ionic radius is a useful one.

Conclusion

The electron transfer model, leading to the formation of ions with electronic structures the same as those of noble gases, therefore accounts for the formulae

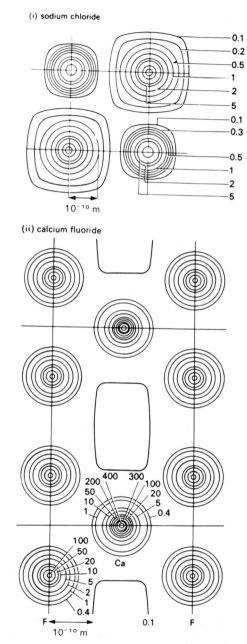

Figure 8.2
Electron density maps for (*i*) sodium chloride and (*ii*) calcium fluoride. Contours are at electrons per 10^{-30} m³.
After WITTE, H. and WOLFEL, E. Reviews of modern physics. **30**, *51–55, 1958.*

of ionic compounds, the charges on them, and thus the non-directional electro-static forces which hold them together in giant lattices.

8.2
COVALENT BONDS

Any model of the bonding in molecules must be able to account for the formula of a molecule, its structure, and the forces which hold the atoms together.

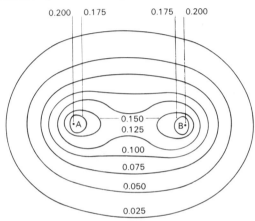

Figure 8.3
Electron density map for the H_2^+ molecule ion. Contours are in electrons per $10^{-30}\,m^3$.
After COULSON, C. A. Proceedings of the Cambridge Philosophical Society, **34**, *1938, 210.*

Figure 8.3 shows an electron density map which was calculated from theory by C. A. Coulson for the simplest possible covalent structure, the H_2^+ molecule ion. This consists of two hydrogen nuclei but only one electron; it thus has a net positive charge. The molecule H_2^+ was chosen for the calculations because of its simplicity.

What do you notice about the contours that is different from the contours in ionic compounds?

What does this tell us about the electron density between the nuclei of the atoms in the molecule?

Figure 8.4 shows an electron density map, determined by X-ray diffraction, for crystals of 4-methoxybenzoic acid. What can be said about the electron density between adjacent atoms in this molecule?

The sets of maps in figures 8.2, 8.3, and 8.4 show that in structures consisting of ions the electron density drops to zero between the ions, and the ions

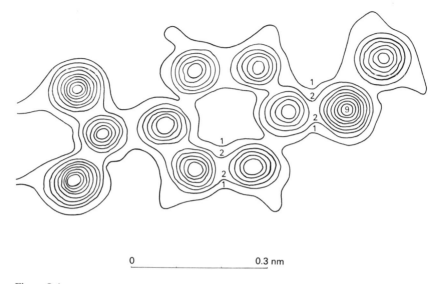

Figure 8.4
Electron density map of 4-methoxybenzoic acid. Contours are in electrons per $10^{-30}\,\text{m}^3$.
Dr J. P. G. Richards, Department of Physics, University College, Cardiff.

are discrete entities; but in molecules there is a substantial electron density at all points along the line joining the centres of two bonded atoms. Thus it seems that in bonds in molecules the electrons are *shared*.

What is it that holds the atoms together in molecules?

Figure 8.5a shows the electron density distribution for the electron cloud in the H_2^+ molecule.

A   B

Figure 8.5a
Charge density distribution for the H_2^+ molecule.

Consider planes through the two nuclei such that the planes are at right angles to the line joining the nuclei, that is, to A and B in figure 8.5b. If a portion, N, of negative charge-cloud is *outside* these planes, as in figure 8.5b, it attracts

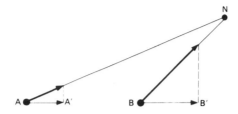

Figure 8.5b
Forces on the nuclei due to a negative charge N not between the nuclei.

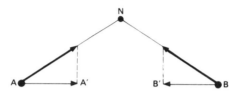

Figure 8.5c
Forces on the nuclei due to a negative charge N between the nuclei.

the nearer nucleus more strongly than the farther one, and tends to separate them. If a portion, N, of negative charge-cloud is *between* the two planes, as in figure 8.5c, it attracts the nuclei towards itself, and thus towards each other (the resolved parts AA' and BB' in figure 8.5c). Since there is very much more negative charge-cloud between the planes than there is outside them, there is a strong net attractive force holding the nuclei together.

The overall effect, therefore, is that the two positive nuclei are bound together by sharing the negative charge-cloud. This arrangement leads to a lower potential energy than if the electron charge-cloud were not shared; and a lower potential energy results in greater stability.

A similar situation exists for the neutral hydrogen molecule, H_2; in this instance, however, the two electrons are shared, and the electron density between the nuclei is greater. The binding is thus stronger than in the instance of H_2^+.

These effects are seen in the separation of the nuclei, and in the bond dissociation energies for H_2^+ and H_2, both of which are given in table 8.1.

	Structure	Internuclear distance/nm	Bond dissociation energy/kJ mol^{-1}
Hydrogen molecule ion, H_2^+	$H \cdot H^+$	0.104	257
Hydrogen molecule H_2	$H : H$	0.074	436

Table 8.1
Data for the hydrogen molecule ion, and the hydrogen molecule.

If the electron density of the hydrogen molecule had been displayed in three dimensions, instead of the two-dimensional representation given in figure 8.5a, it would be seen that the electron density is symmetrical about the axis joining the hydrogen atoms. Bonds of this type are known as σ-bonds (sigma bonds). An important example of a σ-bond is the single bond between carbon atoms, as in the ethane molecule, CH_3—CH_3 (figure 8.6). We shall be considering the carbon–carbon σ-bond in Topic 9.

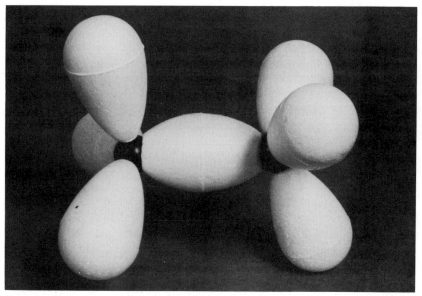

Figure 8.6
A PEEL model showing the electron density distribution in ethane, CH_3—CH_3.
Model, Griffin & George Ltd; photograph, University of Bristol, Faculty of Arts Photographic Unit.

Have another look at figure 8.4. You will see that between adjacent atoms there is a substantial electron density, amounting to between 1 and 2 electrons per unit of volume*, and in some instances between 2 and 3 electrons per unit of volume. These adjacent atoms are bound together by the attraction which the shared electron charge-cloud exerts upon the nuclei on either side of it.

Stoicheiometry, and electron sharing

Covalent bonding exists between atoms when electrons are shared, usually in

* The unit of volume used is $10^{-30}\,m^3$.

pairs. In many cases, the number of atoms involved is such as to enable a noble gas electron structure to be built up around each atom. Figure 8.7 shows how this is done in the case of methane, CH_4.

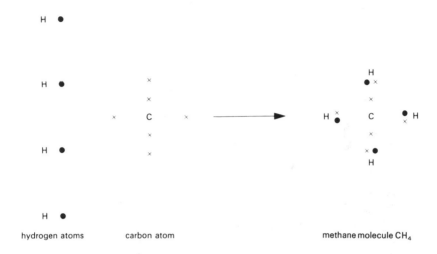

hydrogen atoms   carbon atom   methane molecule CH_4

Figure 8.7
The formation of covalent bonds.

The hydrogen atoms have a share in the electrons from the carbon atom, thus acquiring helium structures; and the carbon atom has acquired the neon structure by sharing electrons from the hydrogen atoms.

Figure 8.8 overleaf shows the formulae and shapes of some molecules in the period lithium to neon. The shapes of the molecules are known from electron diffraction and other physical studies. Copy these into your notebook and then, below the diagram showing the shape of each molecule, draw a dot-and-cross diagram of its electronic configuration. The central atoms of the molecules of beryllium chloride and boron trifluoride do not acquire the electronic structure of a noble gas, but the remainder do.

When atoms of non-metals are joined together it is in general by covalent bonds.

Now try to draw dot-and-cross diagrams showing the electronic configurations in the molecules of hydrogen, H_2; chlorine, Cl_2; hydrogen chloride, HCl; chloromethane, CH_3Cl; methanol, CH_3OH; ethane, CH_3CH_3; ethanol, CH_3CH_2OH; ethene, $CH_2\!=\!CH_2$; ethyne, $CH\!\equiv\!CH$; oxygen, O_2; propane, C_3H_8; and hydrogen cyanide, HCN.

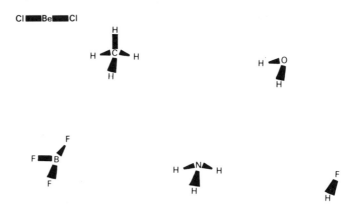

Figure 8.8
The shapes of some molecules in the period lithium to neon.

Molecular shapes, and electron distributions

Single bonds

Refer to the shape of the molecules shown in figure 8.8 and to their electronic configurations. Remember that a covalent bond consists of an internuclear negative charge-cloud. Why do you think that the molecules have the shapes that they do?

What do you notice about the spatial arrangements of the bonds in the molecules of $BeCl_2$, BF_3, and CH_4? What does this suggest about the interaction between electron charge-clouds?

Why are the molecules of ammonia and water not planar and linear respectively?

The bond angles in methane, ammonia, and water molecules are given in table 8.2.

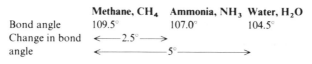

	Methane, CH_4	Ammonia, NH_3	Water, H_2O
Bond angle	109.5°	107.0°	104.5°
Change in bond angle	←——2.5°——→		
	←————————5°————————→		

Table 8.2
Bond angles in some hydrides.

What do these figures suggest?

Multiple bonds

From your consideration of the shapes of molecules containing single bonds, and the changes in the bond angles from one hydride to another as shown in table 8.2, you should have come to the following conclusions:

a Pairs of electrons try to get as far away from each other as they can; this results in a tetrahedral distribution of electron pairs.

b 'Lone' pairs of electrons, that is, pairs of electrons not shared between two atoms, repel one another more strongly than shared pairs do. This causes some distortion of the tetrahedral arrangement.

If you did not come to these conclusions you may like to go back over the evidence to see how well they account for the observations.

On the basis of these ideas, which were developed for single bonds, can you suggest a shape for the molecule of ethene, $CH_2{=}CH_2$? Look again at the dot-and-cross diagrams you drew, based on figure 8.8.

A double bond, as in ethene, consists of two pairs of shared electrons. However, it does not follow that a double bond can be thought of as two single bonds, although it might seem so from our usual method of representing double bonds (figure 8.9).

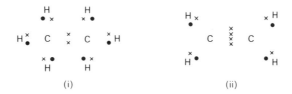

(i)   (ii)

Figure 8.9
Representations of (*i*) ethane, and (*ii*) ethene.

Evidence from both chemical properties and structural studies suggests that the electron density in a double bond is *asymmetric* about the axis joining the two nuclei. In the case of a carbon–carbon double bond the electron density distribution corresponds to a σ-bond (sigma bond) plus what is described as a π-*bond* (pi bond). The π-bond consists of two electron clouds, as shown in figure 8.10, which are not symmetrical about the axis joining the carbon nuclei. A σ-bond was shown previously in figure 8.6.

Using the same approach, suggest a shape for the molecule of ethyne, CH≡CH.

The ideas developed above do enable one to predict correctly the shapes of a surprisingly large number of molecules and other structures; but they are subject to some limitations. Some of these will be discussed later.

Figure 8.10
A PEEL model showing the electron density distribution in ethene, $CH_2{=}CH_2$.
Model, Griffin & George Ltd; photograph, University of Bristol, Faculty of Arts Photographic Unit.

Bond lengths, and bond energies

The greater electron density between nuclei joined by multiple bonds causes a greater force of attraction between the nuclei and is reflected in shorter bond lengths and greater bond energies. Examine the figures given in table 8.3 to confirm this. Are the bond energies of multiple bonds simple multiples of the bond energies of single bonds?

Bond	Compound(s)	Bond length/nm	Bond energy /kJ mol^{-1}
C—C	hydrocarbons	0.154	347
C=C	ethene	0.134	598
C≡C	ethyne	0.121	837
C—N	amines	0.147	286
C=N	oximes	0.132	615
C≡N	hydrogen cyanide	0.116	891
C—O	ethers	0.143	358
C=O	ketones	0.122	749
C≡O	carbon monoxide	0.113	1077

Table 8.3
Bond lengths and bond energies.

Dative covalency

The two electrons which form a covalent bond between two atoms do not necessarily have to come one from each atom; both may originate from one of the atoms.

Look back at the electron configuration which you drew for the compound BF_3. The electrons around the boron atom do not equal the number corresponding to a noble gas; there are two short. Refer also to the electron structure of ammonia; of the eight electrons around the nitrogen atom, two are not shared with any other atom.

Ammonia gas and boron trifluoride gas react readily to give a white solid with the composition NH_3BF_3. This can be interpreted in terms of electron sharing as follows:

$$
\text{H}\overset{\text{x}}{\underset{\text{x}}{\overset{\text{o}}{\text{N}}}}\overset{\text{H}}{\underset{\text{H}}{}} + \text{B}\overset{\text{F}}{\underset{\text{F}}{\text{F}}} \longrightarrow \text{H}\overset{\text{x}}{\underset{\text{x}}{\overset{\text{o}}{\text{N}}}}\overset{\text{H}}{\underset{\text{H}}{}}\text{B}\overset{\text{F}}{\underset{\text{F}}{\text{F}}}
$$

or, using single lines to represent pairs of shared electrons:

$$
\begin{array}{ccc}
\text{H} & \text{F} & \\
| & | & \\
\text{H—N} & + & \text{B—F} \longrightarrow \text{H—N}{\to}\text{B—F} \\
| & | & \\
\text{H} & \text{F} &
\end{array}
$$

Since one pair of the shared electrons has come from one atom the bonding is sometimes known as dative covalency, and the bond is indicated by an arrow $\to$. But the bonding is still covalent and can be written

$$
\begin{array}{cc}
\text{H} & \text{F} \\
| & | \\
\text{H—N—B—F} \\
| & | \\
\text{H} & \text{F}
\end{array}
$$

Another example of dative covalency occurs in the ammonium ion, NH_4^+. This is formed by the combination of a hydrogen ion with an ammonia molecule.

As the next diagram shows, in the ammonium ion the hydrogen atoms each have a share in 2 electrons, giving a helium structure; and the nitrogen atom has a share in 8 electrons, giving a neon structure. The ion has an overall charge

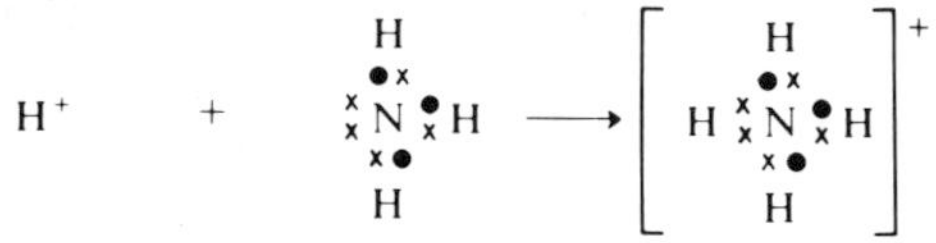

$$H^+ \quad + \quad \overset{\displaystyle H}{\underset{\displaystyle H}{\overset{\bullet\,x}{\underset{x\,\bullet}{{}_x^x N {}_x^\bullet}}}} H \longrightarrow \left[H \overset{\displaystyle H}{\underset{\displaystyle H}{\overset{\bullet\,x}{\underset{x\,\bullet}{{}_x^x N {}_x^\bullet}}}} H \right]^+$$

hydrogen ion   ammonia   ammonium ion
(no electrons)   molecule

of $+1$, originating from the hydrogen ion; it is distributed all over the ion, and is not located on any particular atom.

The sequence shown above leads to a representation of the ammonium ion as in **a** below:

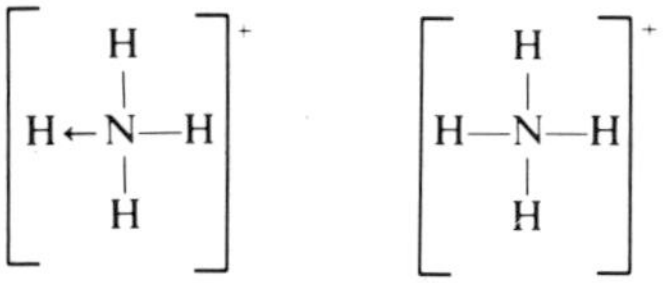

$$\left[\begin{array}{c} H \\ | \\ H \leftarrow N - H \\ | \\ H \end{array} \right]^+ \qquad \left[\begin{array}{c} H \\ | \\ H - N - H \\ | \\ H \end{array} \right]^+$$

a   **b**

But as all the N—H bonds have the same length, and the hydrogen atoms are indistinguishable, a better representation is that in **b**.

The carbon monoxide molecule and the nitric acid molecule also have dative covalent bonds.

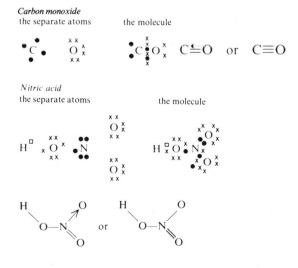

Carbon monoxide
the separate atoms   the molecule

Nitric acid
the separate atoms   the molecule

The bonding in the two NO bonds is not, however, to be taken as different; this point is discussed later (page 246).

How is an oxonium ion, H_3O^+, formed from a water molecule? Draw an electronic structure for the ion. How do you think the positive charge is distributed?

Would it be possible, in terms of electrons, to form an ion H_4O^{2+}? Try to draw such a structure. Have you ever heard of such an ion? Suggest a reason for your answer.

8.3
INTERMEDIATE TYPES OF BONDS

Are bonds either ionic or covalent with no intermediate situation? Or are pure ionic and pure covalent bonds the extreme types, with a complete range of intermediate situations existing in between?

Polarization of ions

Refer back to Topic 6, table 6.1 (page 165), where theoretical values of lattice energies were compared with experimentally determined values. The theoretical values were obtained on the assumption that the ions were spheres, and that transfer of charge had taken place by complete units (that is, electrons). The agreement for the sodium and potassium halides was within 1 per cent, and therefore it appears that the assumption was a reasonable one in these instances.

Table 8.4 gives the values for some other compounds.

| Compound | Structure | Lattice energy/$kJ\,mol^{-1}$ | | |
		Calculated	Experimental	Difference
AgF	NaCl	-920	-958	38
AgCl	NaCl	-833	-905	72
AgBr	NaCl	-816	-891	75
AgI	zinc blende	-778	-889	111
ZnO	wurtzite	-4142	-3971	-171
ZnS	zinc blende	-3427	-3615	188
ZnS	wurtzite	-3414	-3602	188
ZnSe	zinc blende	-3305	-3611	306

Table 8.4
Calculated (Born–Mayer) and experimental (Born–Haber cycle) lattice energies for some compounds.

Work out two or three discrepancies as approximate percentages of the experimental values. Do you think that the pure ionic model holds in these instances?

Spectroscopic studies of the vapours of the alkali metal halides show that these contain diatomic molecules, MX, and that the internuclear distance in these molecules is less than in the corresponding ionic solid. This is illustrated by lithium bromide and lithium iodide. Values for these compounds are given in table 8.5.

	Internuclear separation/nm	
Halide	Crystal	Vapour
LiBr	0.275	0.217
LiI	0.300	0.239

Table 8.5
Internuclear distances in lithium halides.

The shortening of the separation implies stronger bonding than in the crystal, and this can only be achieved by a higher concentration of electrons between the two nuclei. This implies a distortion of the electron cloud of one ion, or both, from a spherical distribution.

Figure 8.11 illustrates the effect.

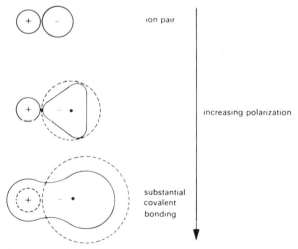

Figure 8.11
Increasing polarization of a negative ion by a positive ion.

The polarization of the ion represents the start of transition from ionic bonding to covalent bonding.

What factors might affect the extent to which an electron charge-cloud around an ion is distorted?

What type of ion would be best at causing distortion? What type of ion would be most easily distorted? Consider the sizes of ions, and the number of charges on ions. Make a list of the various situations.

Is there any evidence in table 8.4 to support your ideas?

Bond polarization, and electronegativity

Two electrons shared between two atoms constitute a covalent bond between these atoms. It is reasonable to ask whether the electrons are always shared equally between the two atoms, or whether some elements are more 'electron-attractive' than others. It is found that elements do differ considerably in their electron-attractiveness. The term used for this is *electronegativity*.

The electronegativity of an atom represents the power of an atom in a molecule to attract electrons to itself.

Many attempts have been made to allot numerical values for the electro-negativities of the elements, but so far no wholly satisfactory method has been devised; each method suffers from shortcomings. But whatever numerical scale is used, the trends in the values of the electronegativities of the elements in the Periodic Table are clear.

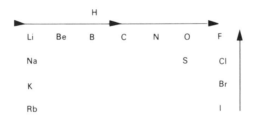

Figure 8.12
Trends in electronegativity in the Periodic Table.

It can be seen, from the trends, that the most electronegative element is fluorine; chlorine, oxygen, and nitrogen are also very electronegative.

If two atoms bonded covalently are atoms of the same element, then the attractions of their nuclei for the bonding electrons are the same, and the bonding electrons will be shared equally between them. But if the atoms are not of the same element, the two nuclei exert different degrees of attractive force on the bonding electrons, and these electrons are displaced towards one atom.

The next diagram illustrates this.

H **:** H   H ×Cl× (with × pairs above, below and · pair)

equal sharing   unequal sharing

(dot-and-cross diagrams: H₂, HCl, Cl₂, and CH₃CH₂Cl)

This unequal sharing of electrons is known as *bond polarization*. It represents the departure of the bond from being purely covalent, and it introduces some ionic character into the bond.

Thus the polarization of ions represents the existence of some covalent character in the ionic bonding; and the polarization of a covalent bond represents the existence of some ionic character in the covalent bond.

One important conclusion from this section so far is that wholly ionic and wholly covalent bonds are extreme types, and examples occur over the whole range of intermediate types: bonds can be partially ionic and partially covalent in character.

Electronegativity and polar molecules

Using the dot-and-cross diagrams above for the HCl molecule and the CH_3CH_2Cl molecule as a starting-point, draw representations of the electron charge-cloud distributions in the two molecules. Do this both for the polarized bonds and the lone pairs of electrons. Superimpose on these drawings a series of + signs to indicate the positions of the atomic nuclei. Now consider whether you think the 'centre of gravity' of the positive charges, and the 'centre of gravity' of the negative charges coincide.

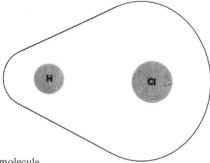

Figure 8.13
The shape of the HCl molecule.

Figure 8.13 represents schematically the shape of the HCl molecule. Copy the figure into your notebook, and draw in by means of a + and a − the relative positions of the centre of positive charge and the centre of negative charge.

What implications do you think this has for the properties of the molecule?

Suggest relative positions for the centre of positive charge and the centre of negative charge in each of the molecules shown in figure 8.14.

Figure 8.14
The molecules of $CHCl_3$ and CCl_4.

What elements in addition to chlorine would be likely to produce these effects?

In asymmetric molecules such as HCl and CH_3CH_2Cl the centre of positive charge does not coincide with the centre of negative charge, and a permanent dipole results. Such molecules are said to be *polar*. Highly electronegative elements such as F, Cl, O, and N cause polarity in molecules. They do so partly by virtue of the bond polarization which they produce, and partly by virtue of their lone pairs of electrons.

Polarity in molecules has important effects on the physical and chemical properties of the substances, and on the mechanisms by which they undergo reaction. These matters will be discussed more fully elsewhere.

Look back to the charge-cloud diagrams which you drew in your notebook for the molecules shown in figure 8.8. Decide which of these molecules are polar and which are non-polar, and label them as such in your drawings.

8.4
DELOCALIZATION OF ELECTRONS

Do single, and double (and triple) covalent bonds always represent the electron distribution adequately when electrons are shared?

We shall examine structural evidence and thermochemical evidence for some compounds, with a view to obtaining an answer to this question.

Nitric acid, and the nitrate ion

Refer back to the electron diagram which you drew earlier for the structure of nitric acid, and answer the following questions in your notebook.

Would any other arrangements of the bonds have done equally well? If so, what are they?

In a range of other compounds the average $N{=}O$ bond length is 0.114 nm and the average $N{-}O$ bond length is 0.136 nm. On the basis of your diagram what would you expect the bond lengths to be in nitric acid?

Electron diffraction and microwave spectroscopic studies of nitric acid vapour show the molecule to have the structure shown below.

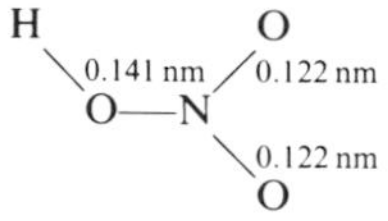

bond lengths 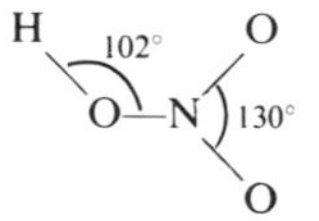   bond angles

Are the bond lengths what you expected?

Draw an electron diagram for the nitrate ion, NO_3^-; and then draw a bond diagram using a line — to represent a single covalent bond and an arrow $\rightarrow$ to represent a dative covalency. The nitrate ion possesses one electron which has been transferred to it from an atom which is now a positive ion.

Is there any other way in which the bonds could be arranged?

What would you expect the bond lengths in the nitrate ion to be?

The structure of the nitrate ion has been found to be as shown below.

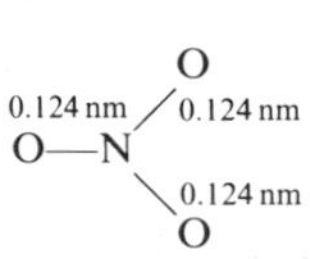

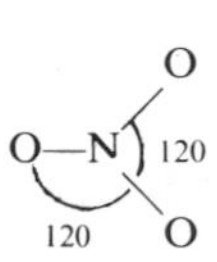

What can be concluded about the nature of the bonds in the nitrate ion?

Methanoic acid, and the methanoate ion

The structural formula of methanoic acid is

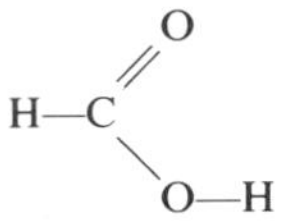

$$H-C\overset{\displaystyle\nearrow O}{\underset{\displaystyle\searrow O-H}{}}$$

On the basis of this diagram, what would you expect the bond lengths to be in methanoic acid?

An electron diffraction study of methanoic acid vapour shows it to have the structure as now shown.

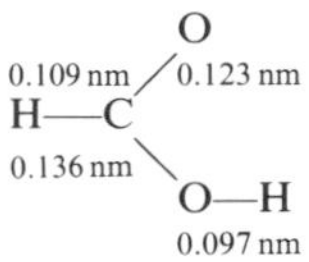

$$\begin{array}{c}\text{O}\\0.109\,\text{nm}\diagup 0.123\,\text{nm}\\ H-C\\0.136\,\text{nm}\diagdown\\ \quad O-H\\ 0.097\,\text{nm}\end{array} \qquad \begin{array}{c}\text{O}\\ H-C \enspace 122^\circ\\ OH\end{array}$$

bond lengths 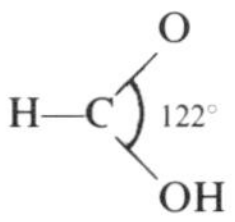   bond angles

Are the bond lengths what you expected?

Earlier in this Topic it was seen that the shapes of molecules could be explained by supposing that pairs of electrons repel one another; thus bonds, and lone pairs of electrons, tend to get as far away from one another as possible.

Are the bond angles in methanoic acid what you would expect from this theory?

Would a small departure from the simple predicted angle seem a likely situation?

Sodium methanoate has the formula $HCO_2^- Na^+$, and that of the methanoate ion is HCO_2^-. Draw an electron structure for the methanoate ion; then draw a diagram using — for a single covalent bond and = for a double covalent bond.

Are alternative diagrammatic arrangements of the bonds possible? What sort of length would you expect the C—O distance to be?

X-ray diffraction studies of sodium methanoate show the methanoate ion to have the structure given below.

O
/ 0.127 nm
H—C
\ 0.127 nm
O

bond lengths

O
/
H—C) 124°
\
O

bond angles

Are the bond lengths within the limits which you expected?

What can you say about the nature of the C—O bonds in the methanoate ion?

Use the evidence from nitric acid, the nitrate ion, methanoic acid, and the methanoate ion. What can be said about the likely bond lengths in bond systems where single and double bonds may be represented diagrammatically by two or more interchangeable arrangements?

Can the structures of such compounds be described adequately in terms of the types of bonds which you have so far studied?

Where equivalent alternative bond structures can be drawn for a compound the actual situation is neither of these; the bond lengths in these situations often prove to be equal, and the available electrons must therefore be distributed equally among the atoms concerned. It is believed that each atom is bonded to the next by two electrons between the nuclei, forming a single covalent bond; and that the remaining available electrons are distributed as charge-clouds above and below all the atoms concerned. These electron charge-clouds are not associated with any particular atom but are mobile over the whole atomic system; they are thus known as *delocalized electrons*. Representations of the situation in nitric acid and the nitrate ion are given in figure 8.15.

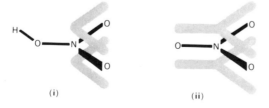

(i)   (ii)

Figure 8.15
Delocalization of electrons in (*i*) nitric acid, and (*ii*) the nitrate ion.

Draw a representation of the electron distribution in the methanoate ion.

Benzene

The organic compound benzene, C_6H_6, has a molecular structure in which delocalization of electrons occurs. The evidence for this is discussed in Topic 9, starting on page 301, and the chemical properties of benzene are considered at the same time. You will find that these properties are very different from those that might be expected for a compound whose molecules were not delocalized in structure.

8.5
METALLIC BONDS

Three significant properties of metals are their high melting points (as contrasted with most non-metals), their high electrical conductivity, and their high thermal conductivity. Any model of the nature of the bonding in metals must be able to account for these properties. Figure 8.16 shows a representation of a simple model of bonding in a solid metal; it consists of metal ions surrounded by a sea of mobile electrons.

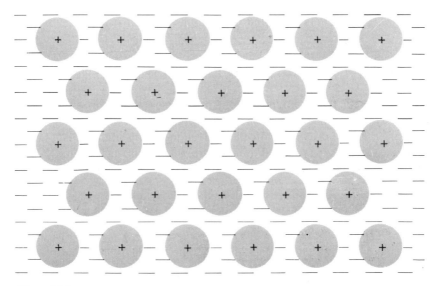

Figure 8.16
A simple representation of bonding in a metal lattice.

The shared electron sea bonds the metal ions tightly into the lattice and confers a relatively high melting point, while the mobile electrons provide a means of conducting electricity and heat. The mobile electrons are another example of delocalization.

This model is an oversimplification, and it cannot account for all of the properties of metals; the more sophisticated models are, however, too advanced to consider in this course.

Figure 8.17 shows an electron density map for aluminium. Are the ions spherical?

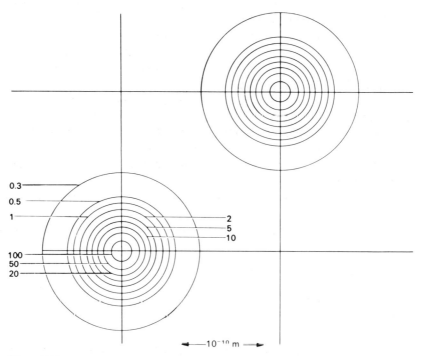

Figure 8.17
Electron density map of aluminium. Contours are in electrons per 10^{-30} m³.
(The average density of electrons between the ions is found to be 0.21 electron per 10^{-30} m³.)
After WITTE, H. and WOLFEL, E. Reviews of modern physics, **30**, *51–55*, 1958.

SUMMARY

At the end of this Topic you should:

1 know how electron transfer, to give ionic bonding, accounts for the stoicheiometry of ionic compounds, and the forces which hold the ions together;

2 understand the meaning of ionic radius and recognize its limitations;

3 understand how the sharing of pairs of electrons between two nuclei leads to a covalent bond;

4 be aware of X-ray diffraction evidence (electron density diagrams) for electron transfer and electron sharing;

5 be able to construct dot-and-cross electron diagrams for ions, and for simple molecules containing single, double, or triple covalent bonds, or dative covalent bonds;

6 understand how bond pair and lone pair electrostatic repulsion accounts for the shapes of simple molecules;

7 know that multiple bonds have a shorter bond length, and a greater bond energy, than single bonds;

8 understand what is meant by polarization of an ion, know the types of ion that are most readily polarized, and appreciate the evidence for polarization of ions provided by the comparison of theoretical and experimental values of lattice energies;

9 understand the term electronegativity, and know the trends in electronegativity in the Periodic Table;

10 understand what is meant by a polar molecule, and appreciate the evidence for the existence of molecules of this kind;

11 recognize that ionic and covalent bonds are extreme types, and that most bonds are intermediate in character;

12 understand what is meant by delocalization, and know something of the characteristics of delocalized structures;

13 know the simple ion lattice and electron sea model of metallic bonding, and recognize that it accounts for the strong bonding in metals, and the good electrical and heat conductivity.

PROBLEMS

* Indicates that the *Book of data* is needed.

1 The following is an extract from a textbook of structural chemistry.

'In all its compounds, nitrogen has four pairs of electrons in its valency shell. According to the numbers of the lone pairs, there are the five possibilities exemplified by the series

A	B	C	D	E
NH_4^+	NH_3	NH_2^-	NH^{2-}	N^{3-}
ammonium ion	ammonia	amide ion	imide ion	nitride ion

The last three, the NH_2^-, NH^{2-} and N^{3-} ions, are found in the salt-like amides, imides, and nitrides of the most electropositive metals.'

i Draw dot-and-cross diagrams of the structures A to E.
ii Sketch the shapes you would expect NH_4^+ ions and NH_3 molecules

to have. Explain the differences, if any, in the HNH bond angles in NH_4^+ and NH_3.

iii Show by means of a sketch the shape you would expect the amide ion to have. Make an estimate of the likely values of the bond angles.

iv The ammonia molecule, NH_3, can form the positive ion NH_4^+. Would you expect methane, CH_4, to form an ion CH_5^+? Give reasons for your answer.

2 The carbonate ion, CO_3^{2-}, has a planar structure as shown in the diagram below.

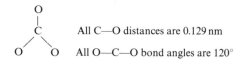

All C—O distances are 0.129 nm

All O—C—O bond angles are 120°

The diagram shows only internuclear separations and shape, not bonding.

a What structures can you draw for this ion using the dot-and-cross method?

*b Select any *one* of the structures you have drawn. What bond lengths does it suggest the ion should have? Do these agree with the observed internuclear separations? If not, how do they differ?

c Suggest a way of explaining the nature of the bonds in the carbonate ion.

d It has been suggested that molecules of carbonic acid, H_2CO_3, are present in low concentration in aqueous solutions of carbon dioxide. How, if at all, would you expect the bond lengths and angles between carbon and oxygen atoms in the carbonic acid molecule to differ from those in the CO_3^{2-} ion?

3 Use the following data for this question.

Bond			Bond length /nm	Bond energy /kJ mol^{-1}
N—N	hydrazine	N_2H_4	0.145	158
N=N	azomethane	CH_3—N=N—CH_3	0.120	410 $\{$ bond dissoci-
N≡N	nitrogen	N_2	0.110	945 $\{$ ation energy
C≡O	carbon monoxide	CO	0.113	1077 (bond energy)

a Do these figures support the statement in the first paragraph following the heading 'Bond lengths and bond energies' on page 238?

b Compare by means of dot-and-cross diagrams the isoelectronic molecules CO and N_2 (isoelectronic means having the same number of electrons). In view of what you know about their relative tendencies to react chemically, is there anything about their bond dissociation energies which surprises you? If so, what?

c Draw (i) a dot-and-cross diagram for the structure of hydrazine, and (ii) a sketch of a hydrazine molecule showing the bond angles you would expect to find in the molecule.

d Draw (i) a dot-and-cross diagram for the structure of azomethane, and (ii) a sketch of one structure of azomethane showing the bond angles you would expect to find in the molecule. Can you then sketch a second structure showing an azomethane molecule with the same bonding but a different shape?

4 Sulphur forms a chloride, SCl_2, in the gas phase. Draw a diagram of the molecule showing the shape you would expect it to have, and indicating the approximate value you would expect for the bond angle.

5 Use the general trends in electronegativity evident from figure 8.12, to explain the following:

a Sodium hydride has a structure which contains the ions Na^+ and H^-.

b Carbon hydride (methane) has a covalent molecular structure and the electrons are evenly shared in the bonds between the carbon and hydrogen atoms in the methane molecules.

c Chlorine hydride (hydrogen chloride) gas has a covalent molecular structure but the molecule has a dipole

H—Cl

$\delta + \delta -$

d Lithium forms a crystalline fluoride Li^+F^- whereas oxygen forms a gaseous fluoride OF_2.

An introduction to organic chemistry

In this first Topic on organic chemistry, we shall begin by considering some of the reasons for the great diversity of carbon compounds, and some of the rules necessary for naming them. Next, we shall consider the results of experiments on four different types of carbon compounds. This will help us to examine an interpretation of the fundamental ways in which carbon compounds react. We shall also meet the important reactions of these carbon compounds and learn something of their social and industrial importance.

This Topic contains a number of ideas that will be new to you. However, the three remaining Topics on organic chemistry contain a further exploration of these ideas, rather than a large number of additional ones. It is important, therefore, that you study this first Topic carefully before you proceed to Topic 11. The last section is a summary which is designed to help you learn the essential framework of ideas and reactions that have been included.

9.1
THE VARIETY OF MOLECULAR STRUCTURE IN ORGANIC COMPOUNDS

Although the chemistry of carbon can be studied as a part of Group IV of the Periodic Table, carbon also has a special chemistry of its own, a chemistry that has been able to flourish in the conditions on this planet. If we compare carbon compounds with those of silicon, as silicon is the nearest neighbour to carbon in Group IV of the Periodic Table, two major differences can be identified:

   1 carbon compounds frequently contain long chains of carbon atoms,

$$
\begin{array}{cccc}
| & | & | & | \\
-C & -C & -C & -C-, \text{ whereas} \\
| & | & | & |
\end{array}
$$

silicon compounds commonly contain chains made up of silicon and oxygen,

$$
\begin{array}{cc}
 & | & | \\
-O & -Si & -O & -Si- \\
 & | & | \\
\end{array}
$$

   2 carbon atoms never bond to more than four other atoms.

   At a simple level the first of these differences between carbon and silicon chemistry can be interpreted as due to differences in bond energies (look again at Topic 6 if necessary).

$$E(\text{C}-\text{C}) \qquad 347\,\text{kJ}\,\text{mol}^{-1}$$

$E(C—H)$ $413 \, \text{kJ mol}^{-1}$
$E(C—O)$ $358 \, \text{kJ mol}^{-1}$

For single bonds these are all high values of similar magnitude, which means that the bonds can be classified as strong bonds of about the same strength. Thus there will be no strong tendency for a C—C bond or a C—H bond to be replaced by a C—O bond (under standard conditions).

Now look at the corresponding values for silicon:

$E(Si—Si)$ $226 \, \text{kJ mol}^{-1}$
$E(Si—H)$ $318 \, \text{kJ mol}^{-1}$
$E(Si—O)$ $466 \, \text{kJ mol}^{-1}$

The Si—Si bond can be classified as a weak bond with a strong tendency to be replaced by Si—O bonds. So, on the Earth, silicon is found in silicate rocks while carbon can be found in a rich variety of carbon–hydrogen compounds as well as in rocks containing metal carbonates.

The second difference can be described as a difference in possible electron arrangements in Group IV elements. When carbon atoms have formed four bonds, they have eight electrons in the second electron shell, and this shell is incapable of further expansion. This restricts the possibilities of further chemical attack.

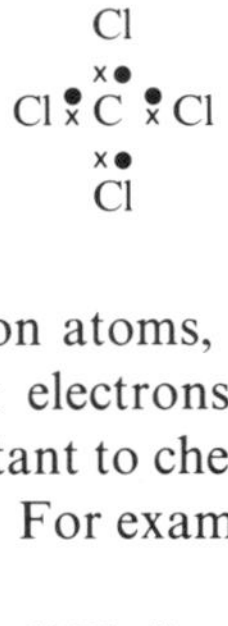

$$
\begin{array}{c}
\text{Cl} \\
\overset{\times\,\bullet}{\text{Cl}} \overset{}{\times}\, \text{C} \,\overset{}{\times}\, \text{Cl} \\
\overset{\times\,\bullet}{\text{Cl}}
\end{array}
$$

Silicon atoms, however, can form an outermost electron shell with more than eight electrons, so compounds such as silicon tetrachloride, $SiCl_4$, are not resistant to chemical attack.

For example, silicon tetrachloride reacts with water

$$SiCl_4(l) + 3H_2O(l) \longrightarrow H_2SiO_3(aq) + 4HCl(aq)$$

but carbon tetrachloride (tetrachloromethane) resists attack by water (see Topic 18).

However, bond strengths in carbon compounds, and the limitation of carbon to compounds with no more than eight electrons in the outermost shell around the carbon atom, are only part of the reason for the diversity of carbon compounds. They are very diverse: it is claimed that over 5 million organic compounds of carbon have been prepared or identified in chemistry laboratories around the World and the possible number of organic compounds is very much greater than that. In this course, we shall meet numerous compounds with six carbon atoms and several with two hundred or more (not that these formulae have to be learned!). For instance:

$C_6H_3OCl_3$ TCP, an antiseptic

$C_{254}H_{377}N_{65}O_{75}S_6$ insulin, a hormone

In order to build up a picture of how this variety occurs in organic compounds, consider first four compounds of very similar formulae that are used as fuels:

CH_4 methane, found in natural gas

C_2H_6 ethane

C_3H_8 propane, used for industrial bottled gas, as in Calor gas

C_4H_{10} butane, used for bottled gas, as in Camping Gaz

The carbon atoms in the molecules of these compounds form four covalent bonds arranged in a tetrahedral pattern. The carbon atoms are arranged in chains, and each molecular formula differs from the one next to it on the list by a CH_2 unit, as shown below.

These are the ways of representing the formulae of organic compounds.

a Molecular formulae

CH_4 $\qquad$ C_2H_6 $\qquad$ C_3H_8 $\qquad$ C_4H_{10}

b Structural formulae

CH_4 $\qquad$ $CH_3—CH_3$ $\qquad$ $CH_3—CH_2—CH_3$ $\qquad$ $CH_3—CH_2—CH_2—CH_3$

c Displayed formulae

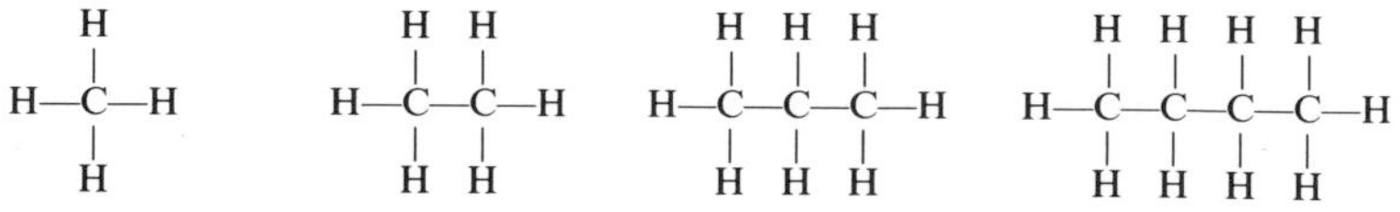

$$
\begin{array}{cccc}
& \text{H} & \text{H H} & \text{H H H} & \text{H H H H} \\
& | & |\ \ | & |\ \ |\ \ | & |\ \ |\ \ |\ \ | \\
\text{H}-&\text{C}-\text{H} & \text{H}-\text{C}-\text{C}-\text{H} & \text{H}-\text{C}-\text{C}-\text{C}-\text{H} & \text{H}-\text{C}-\text{C}-\text{C}-\text{C}-\text{H} \\
& | & |\ \ | & |\ \ |\ \ | & |\ \ |\ \ |\ \ | \\
& \text{H} & \text{H H} & \text{H H H} & \text{H H H H} \\
& \text{methane} & \text{ethane} & \text{propane} & \text{butane}
\end{array}
$$

But there is a further cause of variety in organic compounds, because chemists have found *two* compounds with the formula C_4H_{10}. One has a boiling point of $-1\,°C$ and a standard enthalpy change of combustion of $-2877\,kJ\,mol^{-1}$ while the other boils at $-12\,°C$ and gives $-2869\,kJ\,mol^{-1}$ on combustion. These values are sufficiently different for us to be sure we are dealing with two distinct compounds. The solution to the problem lies in the way the carbon chain is arranged, as shown below.

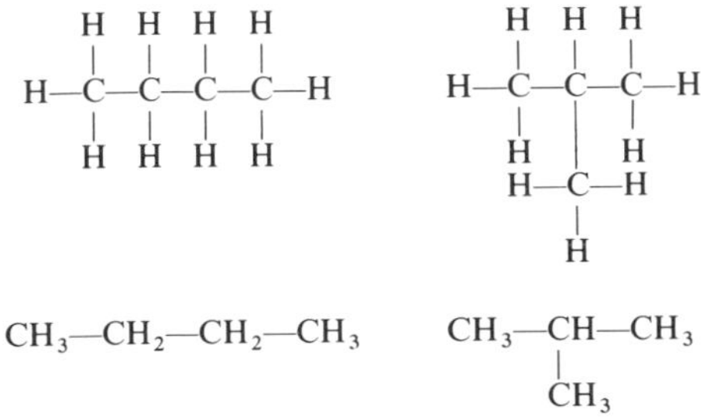

$$
\begin{array}{cccc}
H & H & H & H \\
| & | & | & | \\
H-C-C-C-C-H \\
| & | & | & | \\
H & H & H & H
\end{array}
\qquad
\begin{array}{ccc}
H & H & H \\
| & | & | \\
H-C-C-C-H \\
| & | & | \\
H & H & H \\
 & H-C-H & \\
 & | & \\
 & H &
\end{array}
$$

$$CH_3{-}CH_2{-}CH_2{-}CH_3 \qquad CH_3{-}CH{-}CH_3$$
$$\phantom{CH_3{-}CH_2{-}CH_2{-}CH_3 \qquad CH_3{-}C}\overset{|}{C}H_3$$

butane   2-methylpropane

Compounds that have the same molecular formula but different structures are known as *isomers*.

The existence of isomers makes it necessary to use formulae that show the structure of the molecule, that is, the order in which the atoms are joined together. Full structural formulae, in which every atom is represented separately, are sometimes known as *displayed formulae*. It is often possible to distinguish between isomers by the use of a less detailed *structural formula*. Displayed and structural formulae for the two isomers of C_4H_{10} are shown above.

Now consider the tetrahedral arrangement of groups around the carbon atom more fully. If you make a model of the straight chain C_4H_{10} molecule, you will find that the model can be twisted into a number of different shapes. Is this property of the model shared by the molecule and, if so, can any of the different shapes be described as different compounds? The different shapes cannot correspond to different compounds because, as mentioned previously, chemists have only found properties corresponding to one straight chain compound. It follows, therefore, that the different shapes are all shapes of one compound which must twist just as our model can twist. A glance at a model of the ethane molecule, C_2H_6, shown in figure 9.1, should make this clear. The

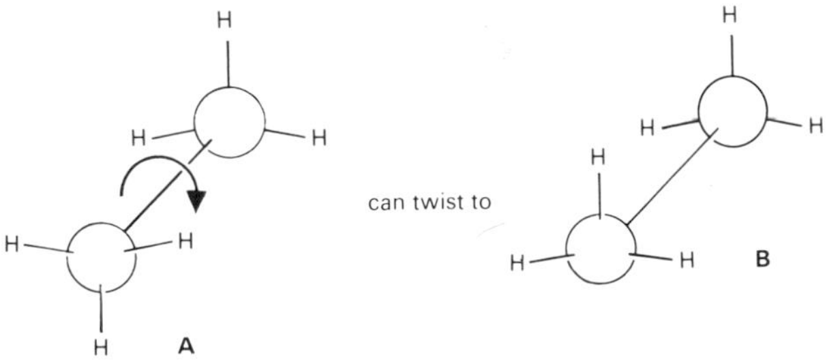

can twist to

Figure 9.1

rotation about the C—C single bond which makes possible the movement seen in the figure is a property of the molecule. So A and B are merely ethane molecules at different stages of a continuous rotation, not isomers of different structures.

Exercise

Make models and write down the structural formulae of all the isomers with molecular formula C_7H_{16}. You should finish with nine isomers.

Some rules for naming organic compounds

The rules for naming compounds have been settled by international agreement through the International Union of Pure and Applied Chemistry. They are usually known as IUPAC rules. The complete set of rules is a very lengthy document and occupies a large book. We need only a few rules to start with, so here we will deal only with compounds of carbon and hydrogen. Some other important groups, the functional groups, will also be mentioned but not discussed in any detail; they need not be learned at this stage.

Compounds containing carbon and hydrogen with only single bonds between the carbon atoms are known as saturated hydrocarbons. They occur as three main types.

a Compounds in which the molecules are made up of straight chains of carbon atoms. The general name for these is *alkanes*. Names for individual compounds all have the ending '-ane'. For example,

$$CH_3—CH_2—CH_2—CH_3 \qquad \text{butane}$$

b Compounds with molecules having branched chains of carbon atoms. These are still known as alkanes, but the rules for the individual names are more complicated. An example is

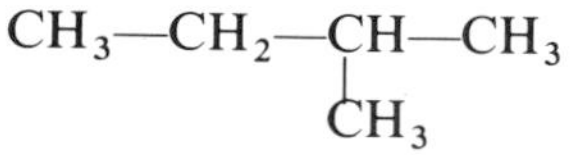

$$CH_3—CH_2—\underset{\underset{\displaystyle CH_3}{|}}{CH}—CH_3 \qquad \text{2-methylbutane}$$

c Compounds with molecules having one or more rings of carbon atoms (to which side chains may be attached). The general name for these is *cyclo-alkanes*. Names for individual compounds contain the prefix 'cyclo-' and have the ending '-ane'. For example,

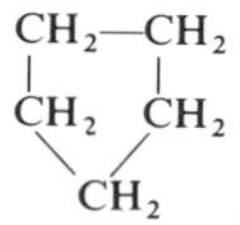

$$
\begin{array}{c}
CH_2—CH_2 \\
|\qquad\ | \\
CH_2\quad CH_2 \\
\diagdown\ \diagup \\
CH_2
\end{array}
\qquad \text{cyclopentane}
$$

a Alkanes containing unbranched carbon atom chains

The names of the first four hydrocarbons in the series, containing 1, 2, 3, and 4 carbon atoms respectively, are methane, ethane, propane, and butane. These do not follow any logical system and must be learned. The rest of the hydrocarbons in the series are named by using a Greek numeral root and the ending '-ane', *e.g.* pentane (5 carbon atoms in an unbranched chain), hexane (6 carbon atoms). The roots are the same as those used in naming geometric figures (pentagon, hexagon, etc.). A list of examples is given below:

Number of carbon atoms in chain	Molecular formula	Name	Number of carbon atoms in chain	Molecular formula	Name
1	CH_4	methane	6	C_6H_{14}	hexane
2	C_2H_6	ethane	7	C_7H_{16}	heptane
3	C_3H_8	propane	8	C_8H_{18}	octane
4	C_4H_{10}	butane	9	C_9H_{20}	nonane
5	C_5H_{12}	pentane	10	$C_{10}H_{22}$	decane
			etc.		

b Alkanes containing branched chains

In order to name these compounds, groups of atoms known as alkyl groups are used. These are derived from hydrocarbons with unbranched carbon chains by removing one hydrogen atom from the end carbon atom of the chain. For example CH_3—CH_2—CH_3 (propane) becomes CH_3—CH_2—CH_2— with one bond unoccupied. Alkyl groups are named from the parent hydrocarbon by substituting the ending '-yl' for the ending '-ane'. Thus, $CH_3CH_2CH_2$—is the propyl group. A list of alkyl groups is given below.

Hydrocarbon	Alkyl group	Formula for alkyl group
methane	methyl	CH_3—
ethane	ethyl	C_2H_5—
propane	propyl	C_3H_7—
butane	butyl	C_4H_9—
pentane	pentyl	C_5H_{11}—
hexane	hexyl	C_6H_{13}—
and so on		

Branched chain hydrocarbons are named by combining names of alkyl groups with the name of an unbranched chain hydrocarbon. The simplest is

$$CH_3—CH—CH_3$$
$$\quad\quad\ |$$
$$\quad\quad CH_3$$

which is called methylpropane

The hydrocarbon name is always derived from the longest continuous chain of carbon atoms in the molecule. The position of the alkyl group forming the side chain is obtained by numbering the carbon atoms in the chain.

$$\overset{1}{C}—\overset{2}{C}—\overset{3}{C}—\overset{4}{C}—\quad \text{etc.}$$

The numbering is done so that the lowest numbers possible are used to indicate the side chain (or chains). Thus

$$CH_3—CH_2—CH_2—CH—CH_3$$
$$\quad\quad\quad\quad\quad\quad\quad\quad |$$
$$\quad\quad\quad\quad\quad\quad\quad\quad CH_3$$

is named 2-methylpentane, not 4-methylpentane which would be obtained by numbering from the other end of the chain. When there is more than one substituent alkyl group of the same kind, the figures indicating the positions of the groups are separated by commas, for example,

$$CH_3—CH—CH—CH_3$$
$$\quad\quad\ |\quad\ |$$
$$\quad\quad CH_3\ CH_3$$

is 2,3-dimethylbutane, and

$$\quad\quad\quad\ CH_3$$
$$\quad\quad\quad\quad |$$
$$CH_3—CH_2—C—CH_3$$
$$\quad\quad\quad\quad |$$
$$\quad\quad\quad\ CH_3$$

is 2,2-dimethylbutane

Different alkyl groups are placed in alphabetical order in the name for a branched chain hydrocarbon, *e.g.*

$$CH_3—CH_2—CH—CH_2—CH—CH_3$$
$$\quad\quad\quad\quad\ |\quad\quad\quad\ |$$
$$\quad\quad\quad\quad CH_2\quad\quad CH_3$$
$$\quad\quad\quad\quad\ |$$
$$\quad\quad\quad\quad CH_3$$

is named 3-ethyl-5-methylhexane

c Alkanes containing a ring of carbon atoms

These are named from the corresponding unbranched chain hydrocarbon by adding the prefix 'cyclo-'. An example is cyclohexane, which can be represented as

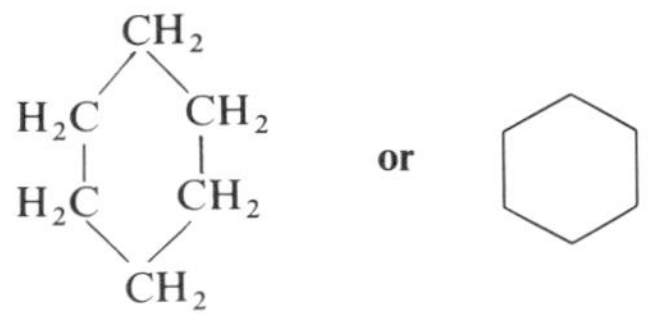

CH₂
H₂C CH₂
H₂C CH₂   or   ⬡
CH₂

All the carbon atoms in an unsubstituted cycloalkane ring are equivalent, so far as substitution is concerned, so that if only one alkyl group is added as a substituent there is no need to number the carbon atoms. Thus methyl-cyclohexane is

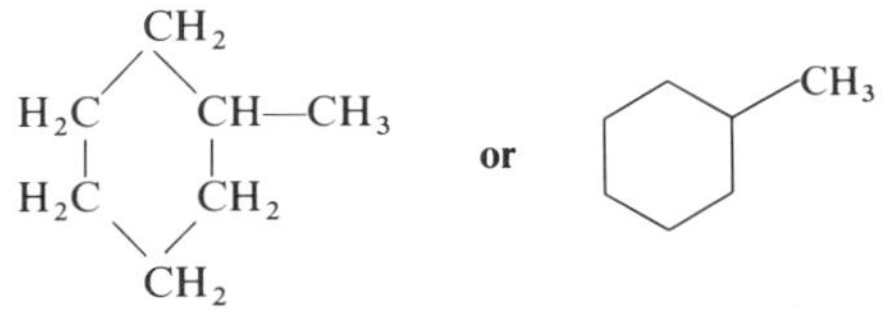

CH₂
H₂C CH—CH₃
H₂C CH₂   or   ⬡—CH₃
CH₂

Names and structures of some functional groups

In this table the structures of the functional groups are printed out in the second column so as to show the atomic linkages. When these structures are repeated in the examples given in the fourth column they are printed on one line only, so as to show this abbreviated method of writing them.

Class of compound	Structure of the functional group	Example of a compound Name	Formula
Alkene	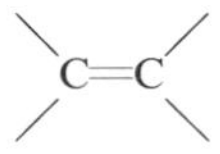 $\diagdown C = C \diagup$	propene	$CH_2{=}CH{-}CH_3$
Arene	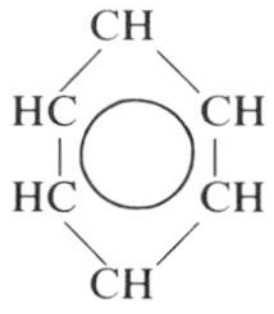 (benzene ring) CH / HC CH / HC CH \ CH	benzene	C_6H_6 or  ⬡
Alcohol	—OH	propanol	$CH_3{-}CH_2{-}CH_2{-}OH$
Amine	—NH₂	propylamine	$CH_3{-}CH_2{-}CH_2{-}NH_2$
Nitrile	—C≡N	propanenitrile	$CH_3{-}CH_2{-}CN$
Halogeno	—Cl etc.	1-chloropropane	$CH_3{-}CH_2{-}CH_2{-}Cl$
Aldehyde	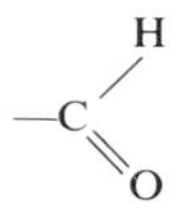 —C(H)=O	propanal	$CH_3{-}CH_2{-}CHO$

Class of compound	Structure of the functional group	Example of a compound Name	Formula
Ketone	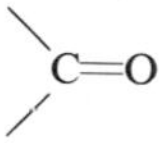 $\diagup\!\!C\!\!=\!\!O\diagdown$	propanone	$CH_3-CO-CH_3$
Carboxylic acid	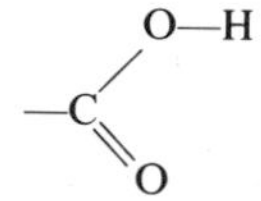 $-C\underset{O}{\overset{O-H}{\big<}}$	propanoic acid	$CH_3-CH_2-CO_2H$
Carboxylate ion	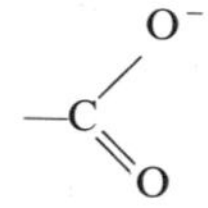 $-C\underset{O}{\overset{O^-}{\big<}}$	sodium propanoate	$CH_3-CH_2-CO_2^-\,Na^+$
Acyl chloride	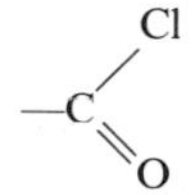 $-C\underset{O}{\overset{Cl}{\big<}}$	propanoyl chloride	CH_3-CH_2-COCl
Acid anhydride	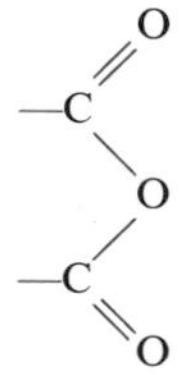 $-C\overset{O}{\underset{O}{\big<}}$ $-C\overset{}{\underset{O}{\big<}}$	propanoic anhydride	$(CH_3-CH_2-CO)_2O$
Amide	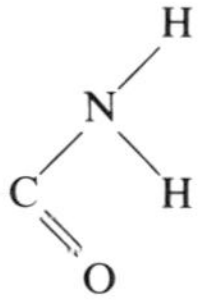 $C\overset{N-H,\,H}{\underset{O}{\big<}}$	propanamide	$CH_3-CH_2-CONH_2$
Nitro compound	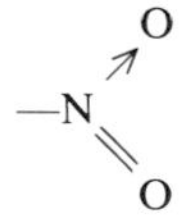 $-N\overset{O}{\underset{O}{\big<}}$	nitrobenzene	$C_6H_5NO_2$ or 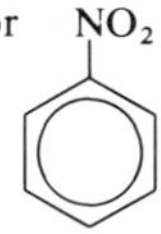 $\quad NO_2$
Sulphonic acid	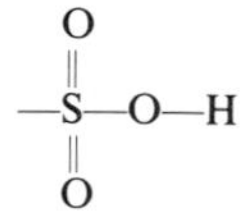 $-\overset{O}{\underset{O}{\overset{\|}{\underset{\|}{S}}}}-O-H$	benzenesulphonic acid	$C_6H_5SO_2OH$ or 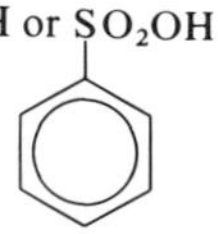 SO_2OH
Ether	$-C-O-C-$	ethoxyethane	$CH_3-CH_2-O-CH_2-CH_3$

9.2
THE ALKANES

The alkanes are saturated hydrocarbons; they have only single bonds in their structure (they are saturated) and are composed only of hydrogen and carbon (hydrocarbons). Alkanes are the major components of crude oil so it is reasonable to claim that they are *the* compounds of the twentieth century: the largest international companies make their living from alkanes and the richest men made their fortunes from alkanes. Countries with crude oil have been able to transform their way of life from that of primitive farmers to that of modern technologists. The money derived from crude oil has enabled governments to build schools, roads, and hospitals, but it has also disrupted traditional ways of life and been used to buy modern war weapons. The price of crude oil, and ensuring a supply of crude oil, have come to dominate modern economic thinking and modern industrial planning. In 1980 consumption of crude oil varied from 30 barrels a year per person in the USA, to $\frac{1}{3}$ barrel a year per person in countries such as India (a barrel contains about $159\,dm^3$). Total World consumption was estimated at 23×10^9 barrels each year and the known reserves still underground were about 650×10^9 barrels. We cannot deduce that the supply of crude oil will run out in 30 years because exploration for new oil fields is usually planned to ensure that newly discovered fields can be brought into production by the

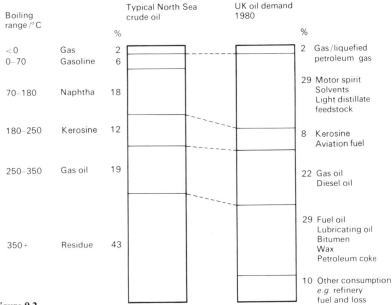

Figure 9.2
Data provided by BP Educational Service.

time they are needed. And that can be a long time: in Alaska from exploration to production took 15 years and in the North Sea the period from exploration to bringing the first oil ashore in the UK was about the same.

Nevertheless, we can be sure that the supply of crude oil is finite so that shortages will develop as oil fields run dry, or as countries restrict output to conserve their source of wealth. It remains to be seen if the World will manage an orderly change to new technologies or whether the rich will outbid the poor in a desperate scramble for oil and whether cars, trains, aeroplanes, and other transport will come to a halt before adequate alternatives are ready. Without care, the future of oil may prove as dark as its past has proved bright.

But how suitable is crude oil for all the potential uses? The composition of a typical barrel of crude oil and the corresponding demand in the market place are illustrated in figure 9.2. Matching supply to demand is the job of the oil refinery. This is a complex task because not only must the demand for alkanes be satisfied but also most of the demand for unsaturated hydrocarbons and arene hydrocarbons. We shall return to this question later in the Topic.

One alkane behaves much like another, so we can describe most of the properties we are concerned with by describing ethane, CH_3—CH_3. The structure of ethane is illustrated in figure 9.3.

Notice that the single bond between the carbon atoms (the σ or sigma bond) is symmetrical, which is consistent with the free rotation about the C—C bond discussed in the previous section.

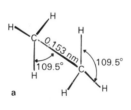
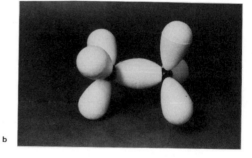

Figure 9.3
The structure of ethane.
a A diagram showing the bond lengths and bond angles.
b A PEEL model of the molecule of ethane showing the distribution of the electrons.
Model, Griffin & George Ltd; photograph, University of Bristol, Faculty of Arts Photographic Unit.

The average bond energies in alkanes are:

E(C—C) $347\,kJ\,mol^{-1}$
E(C—H) $413\,kJ\,mol^{-1}$

If you compare these values with other bond energies in the *Book of data*, they will be seen to be towards the top of the range. What does this suggest about the likely reactivity of the alkanes?

A typical infra-red spectrum of an alkane is shown in figure 9.4.

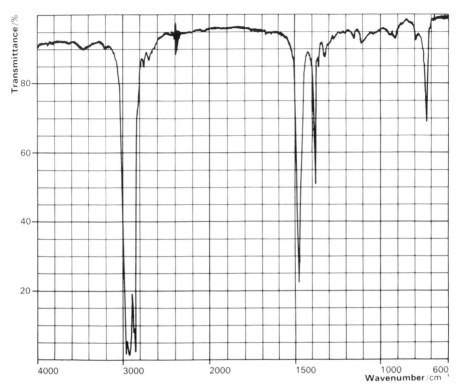

Figure 9.4
The infra-red spectrum of an alkane (decane).

Use the additional charts in Topic 7, section 1, (figures 7.5 and 7.6) to check the assignment of the peaks to the appropriate bonds.

The change in boiling points with increase in number of carbon atoms for the straight-chain alkanes can be seen in figure 9.5 on the next page.

Experiments with alkanes

We shall begin our experimental investigations with a study of some alkanes: hexane (C_6H_{14}), light paraffin (a mixture, containing alkane molecules with about 12 carbon atoms), and poly(ethene) $(CH_3-(CH_2)_n-CH_3)$. *BE CAREFUL*

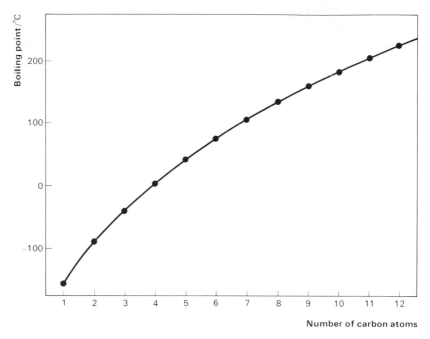

Figure 9.5
The change in boiling points of straight-chain alkanes with increase in number of carbon atoms.

WHEN USING HEXANE: IT IS ABOUT AS VOLATILE AND FLAM-MABLE AS PETROL.

EXPERIMENT 9.2
An investigation of the reactions of some alkanes

Carry out the following tests on a sample of each of the three alkanes. Note your observations in tabular form carefully because you will be doing similar tests on other compounds later in this Topic and you will be expected to compare the behaviour of the different types of compound.

Wear safety glasses throughout this experiment.

Procedure

1 *Combustion* If possible, do this experiment in a fume cupboard. Keep a sample of the liquid alkanes in a test-tube well away from any flame. Dip a combustion spoon into the sample. Set fire to the alkane on the combustion spoon and note the luminosity and sootiness of the flame.

2 *Oxidation* To 0.5 cm³ of the liquid alkane or two or three granules of poly(ethene) in a test-tube add two or three drops of a mixture of equal volumes of 0.01M potassium manganate(VII) and 2M sulphuric acid. Shake the contents and try to tell from any colour change if manganate(VII) oxidizes the alkane.

3 *Action of bromine* To 0.5 cm^3 of the liquid alkane or two or three granules of poly(ethene) in a test-tube add a few drops of a 2% solution of bromine in 1,1,1-trichloroethane. (*TAKE CARE.*)

What, if anything, happens to the colour of the bromine?

4 *Action of bromine in sunlight* To 10 cm^3 of hexane in a test-tube add several drops of 2% bromine in 1,1,1-trichloroethane. Loosely cork the tube and irradiate with sunlight or a photoflood light (*PROTECT YOUR EYES FROM THE LIGHT*). Prepare a second similar tube and leave it in a dark place. Compare the intensity of the bromine colour at intervals.

Is there any evidence of reaction? Can you identify any fumes evolved?
If no fumes are apparent try tipping the contents of the test-tube into
a beaker.

5 *Action of sulphuric acid* Put 1–2 cm^3 of concentrated sulphuric acid in a test-tube held in a rack (*TAKE CARE*). Add 0.5 cm^3 of the liquid alkane or two or three granules of poly(ethene).

Do the substances mix or are there two separate layers in the test-tube?

6 *Action of alkali* To 0.5 cm^3 of the liquid alkane or two or three granules of poly(ethene) in a test-tube add 1–2 cm^3 of 20% potassium hydroxide (*TAKE CARE: IT IS CAUSTIC*) dissolved in ethanol. Mix the liquids by shaking the tube gently. *DO NOT ALLOW THE POTASSIUM HYDROXIDE TO COME INTO CONTACT WITH YOUR SKIN.*

Is there any sign of reaction?

7 *Catalytic cracking* Put light paraffin in a test-tube to the depth of 1–2 cm. Push in some loosely packed ceramic fibre until all the paraffin has been soaked up. Now add 2–3 cm depth of aluminium oxide granules and clamp the test-tube horizontally so that the granules form a layer in the test-tube. Connect the test-tube for collection of gas over water as shown in figure 9.6. Heat the aluminium oxide strongly and continuously but be careful not to melt the rubber stopper, nor to allow the delivery tube to become blocked. The paraffin should get hot enough to evaporate without needing direct heat.

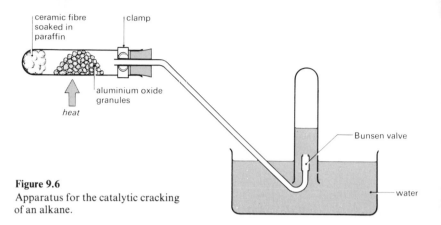

Figure 9.6
Apparatus for the catalytic cracking
of an alkane.

Collect 3 or 4 tubes of gas (discard the first one: why?), and when the delivery of gas slows down, lift the apparatus clear of the water to avoid it being sucked up into the hot test-tube.

Carry out the following two tests on the gas collected.

a *Test for flammability.* Any flame will be more visible if the test-tube is held upside down.

b *Test for reaction with bromine.* Add a few drops of 2% bromine in 1,1,1-trichloroethane. (*TAKE CARE.*)

Is the gaseous product an alkane or can different properties be observed?

An interpretation of the photochemical experiment with alkanes

The lack of positive results when one is doing experiments with alkanes in the laboratory probably seems disappointing but, for most of their uses, the chemical inertness of the alkanes is their greatest asset. Compounds that are non-corrosive to metals (lubricating oils), harmless to our skin (petroleum jelly), and safe in contact with foods (poly(ethene)) are enormously useful to us.

The reaction of the alkanes that we need to consider here is that of hexane with bromine in sunlight. Since the reaction needs light in order to take place (unless the reactants are heated to over 300°C), it is known as a photochemical reaction. It is a general reaction between alkanes and bromine or chlorine. The process by which reaction takes place must depend on the absorption of the energy of the photons that make up the radiation. Other experiments on the reaction of methane with chlorine have shown that the process does not need many photons: many thousands of product molecules are produced for each photon absorbed. So by what process does this reaction occur? To simplify the equations involved we will consider the methane-chlorine reaction as a typical example of what is involved.

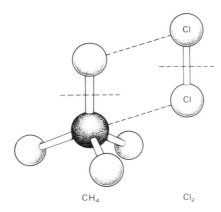

Figure 9.7
A ball-and-stick representation of possible attack by chlorine on methane.

If you look at a ball-and-stick model, such as the one shown in figure 9.7, you may think that the reaction could begin with the simultaneous breaking of the C—H and Cl—Cl bonds, followed by the making of C—Cl and H—Cl bonds. Calculations using the appropriate bond energies show that this would be an exothermic reaction, with an enthalpy change of about $100 \, \text{kJ mol}^{-1}$.

$$CH_4 + Cl_2 \longrightarrow CH_3Cl + HCl; \qquad \Delta H^{\ominus} = -100 \, \text{kJ mol}^{-1}$$

But there are two problems about this proposal. Firstly, the molecules would need to come together with an exact orientation for reaction to happen, and that would be a very rare collision amongst all the collisions occurring. Secondly, the distance between the atoms would have to be reduced to about a bond length and the force of repulsion between the electron clouds would have to be overcome.

How much energy can be provided by the radiation absorbed when the molecules react?

To compute the energy of radiation (E_{mol}) in 1 mole (L) of photons we will use the relationship

$$E_{\text{mol}} = Lh\nu$$

where

$L = 6.02 \times 10^{23} \, \text{mol}^{-1}$ (the Avogadro constant)
$h = 6.62 \times 10^{-37} \, \text{kJ s}$ (the Planck constant)
$\nu = $ frequency of the radiation in hertz (s^{-1})

If we use the data in figure 7.1 an approximate value for the frequency of infra-red radiation is $10^{13}\,s^{-1}$, for visible light is $10^{14}\,s^{-1}$, and for ultra-violet light is $10^{15}\,s^{-1}$. If you carry out this calculation you will find that only ultra-violet photons have sufficient energy to be of interest to us

$$E(\text{ultra-violet photons}) \approx 400\,kJ\,mol^{-1}$$

This is enough energy to break up the chlorine molecule into uncharged chlorine atoms with unpaired electrons, which are given the symbol Cl·

$$Cl_2 \longrightarrow Cl\cdot + Cl\cdot \qquad \Delta H^{\ominus} = +\ 242\ kJ\ mol^{-1}$$

However, it is scarcely enough to break a methane molecule

$$CH_4 \longrightarrow CH_3\cdot + H\cdot \qquad \Delta H^{\ominus} = +\ 435\ kJ\ mol^{-1}$$

and certainly not enough to produce ions.

$$Cl_2 \longrightarrow Cl^+ + Cl^- \qquad \Delta H^{\ominus} \approx +\ 1130\ kJ\ mol^{-1}$$
$$CH_4 \longrightarrow H^+ + CH_3^- \qquad \Delta H^{\ominus} \approx +\ 1700\ kJ\ mol^{-1}$$

We can therefore conclude that the first step in this photochemical reaction is probably the absorption of an ultra-violet photon by a chlorine molecule, resulting in the formation of two chlorine atoms. These chlorine atoms, each with an odd number of electrons, are known as *free radicals*.

Note carefully the difference between the free radical fission of the Cl—Cl bond and the ionic fission of the Cl—Cl bond.

To form a free radical the process is

This is known as *homolytic fission*; 'homo-' is from ancient Greek, meaning 'same'.

To form ions the process is

This is known as *heterolytic fission*, 'hetero-' meaning 'different'.

How does the reaction proceed? When the chlorine radical attacks a methane molecule there seem to be two possibilities for products. Either a hydrogen radical (we have to have a radical product if we start with an odd number of electrons in the reactants) and chloromethane are produced

$$Cl \cdot + CH_4 \longrightarrow H \cdot + CH_3Cl \qquad \Delta H^{\ominus} = +108 \, kJ \, mol^{-1}$$

or a methyl radical and hydrogen chloride are produced.

$$Cl \cdot + CH_4 \longrightarrow CH_3 \cdot + HCl \qquad \Delta H^{\ominus} = +4 \, kJ \, mol^{-1}$$

Thus, on the basis of the energy involved, we can conclude that the second step in the reaction is likely to be the formation of a *methyl free radical*. Similar considerations will lead us to the next step.

$$CH_3 \cdot + Cl_2 \longrightarrow CH_3Cl + Cl \cdot \qquad \Delta H^{\ominus} = -97 \, kJ \, mol^{-1}$$

These two last steps taken together produce the correct end-products, hydrogen chloride and chloromethane, but they also link together in an apparently endless chain for as long as methane and chlorine molecules are available:

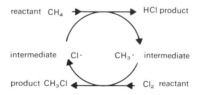

Is this chain likely to go on indefinitely? Well, not really, because to keep going it depends on free radicals colliding only with ordinary molecules. In practice, free radicals are bound to collide with each other, and if you think about the possibilities, you should realize there are three:

$$Cl \cdot + Cl \cdot \longrightarrow Cl_2$$

$$CH_3 \cdot + CH_3 \cdot \longrightarrow CH_3 - CH_3$$

and $$Cl \cdot + CH_3 \cdot \longrightarrow CH_3Cl$$

These collisions will all terminate the chain. Experiments have shown that for each original photon absorbed, on average 10 000 molecules of chloromethane are produced. So how many links are there on average in a chain before the chain is terminated?

Free radical chain reactions are also important in the formation of polymers, such as poly(ethene), and in the combustion of hydrocarbons, especially petrol. We shall return to these topics later on.

Reactions of the alkanes

1 *Combustion* It is the combustion of the alkanes that provides their most important uses:

$$CH_4 + 2O_2 \longrightarrow CO_2 + 2H_2O \qquad \Delta H_c^{\ominus} = -890 \, kJ \, mol^{-1}$$
methane

$$C_4H_{10} + 6\tfrac{1}{2}O_2 \longrightarrow 4CO_2 + 5H_2O \qquad \Delta H_c^{\ominus} = -2877 \, kJ \, mol^{-1}$$
butane

These equations represent very familiar processes because the oxygen for combustion normally comes from the air, while methane is the main component of domestic gas and butane is a component of bottled gas.

Other industrial products, including petrol, jet fuel (kerosine), Diesel oil, paraffin, heating oil, and candlewax are mixtures of saturated and unsaturated hydrocarbons, but the main components are alkanes with appropriate boiling points. In normal use, the alkane fuels have the great advantage of burning with a relatively clean flame and producing non-toxic products. However, if the air supply is restricted, carbon monoxide can be produced:

$$CH_4 + 1\tfrac{1}{2}O_2 \longrightarrow CO + 2H_2O$$

Carbon monoxide is dangerously toxic.

The particular case of the combustion of petrol will be dealt with in more detail later in this Topic.

2 *Photochemical reactions with halogens* The photochemical reaction of chlorine and bromine with alkanes

$$C_6H_{14} + Br_2 \longrightarrow C_6H_{13}Br + HBr$$
hexane   bromohexane

is not a useful method of preparing halogenoalkanes. This is because further reaction takes place, forming a mixture of products. Photochemical and free radical organic reactions are important, however, in other contexts.

This photochemical process is a chain reaction involving free radicals in the following sequence (as explained earlier in this section).

$$Cl_2 \xrightarrow{hv} 2Cl\cdot \qquad\qquad \text{chain initiation}$$

$$\left.\begin{array}{l} CH_4 + Cl\cdot \longrightarrow HCl + CH_3\cdot \\ Cl_2 + CH_3\cdot \longrightarrow CH_3Cl + Cl\cdot \end{array}\right\} \text{chain propagation}$$

$$\left.\begin{array}{l} Cl\cdot + Cl\cdot \longrightarrow Cl_2 \\ CH_3\cdot + CH_3\cdot \longrightarrow CH_3{-}CH_3 \\ Cl\cdot + CH_3\cdot \longrightarrow CH_3Cl \end{array}\right\} \text{chain termination}$$

The overall reaction is

$$CH_4 + Cl_2 \longrightarrow CH_3Cl + HCl$$

This is followed by reaction with more chlorine in further photochemical chain reactions producing

$$CH_3Cl + Cl_2 \longrightarrow CH_2Cl_2 + HCl$$

$$CH_2Cl_2 + Cl_2 \longrightarrow CHCl_3 + HCl$$

$$CHCl_3 + Cl_2 \longrightarrow CCl_4 + HCl$$

3 *Catalytic cracking* This is an important process in the petrochemical industry where much of the fraction from the distillation of crude oil with a boiling range 200–300 °C (C_{10} to C_{20} alkanes) is heated to 500 °C in the presence of a silica–alumina catalyst to produce unsaturated hydrocarbons, alkenes, and short-chain alkanes useful for petrol. The high boiling residues are used as fuel oil.

9.3
THE HALOGENOALKANES

If hydrogen atoms in alkanes are replaced by halogen atoms, we have compounds of the type known as halogenoalkanes. These are named by using the name for the alkane from which they are derived and adding 'chloro', 'bromo' or 'iodo'. For example,

CH_3Cl chloromethane

$CH_3{-}CH_2Br$ bromoethane

Two halogenoalkanes can be derived from propane. They are distinguished by numbering the carbon atoms as for the branched chain alkanes.

$$CH_3—CH_2—CH_2Cl$$ 1-chloropropane

$$CH_3—CHCl—CH_3$$ 2-chloropropane

Halogenoalkanes with side chains are named in the same way as the corresponding alkanes

$$CH_3—CH—CH_2Cl$$ 1-chloro-2-methylpropane
$$\qquad\ \ |$$
$$\qquad\ \ CH_3$$

$$CH_3—CCl—CH_3$$ 2-chloro-2-methylpropane
$$\qquad\ |$$
$$\qquad\ CH_3$$

For halogenoalkanes with more than one halogen atom the full name of the alkane is used, preceded by the number of the carbon atom to which the halogen atoms are attached, with 'di', 'tri', etc., to indicate the total number of halogen atoms. For example

$$CH_2Br—CH_2Br$$ 1,2-dibromoethane

In the whole field of naturally occurring materials, there are practically no halogen compounds. Thus, almost all of them must be produced synthetically. Those few which do occur in nature are found in rather obscure situations. Examples include the iodine compound thyroxine, a hormone produced by the thyroid gland, a shortage of which causes goitre and cretinism; the bromine compound Tyrian purple, present in the viscera (gut) of a type of sea-snail called *Murex brandaris*, which was extracted and used by the Romans for dyeing their statesmen's robes; and the chlorine compound chloromethane, produced by some marine algae.

Owing to the considerable reactivity of the halogen atoms in halogeno-alkanes, many of these compounds are manufactured as 'intermediates', that is, for conversion into other substances. There is also a range of organic halogen compounds which are important products of industry, for example,

CF_2Cl_2 a propellant in aerosol cans

$(CH_2CHCl)_n$ the polymer PVC, poly(chloroethene)

$CHClBr—CF_3$ an anaesthetic, Fluothane (see the Background reading, page 283)

$C_6H_6Cl_6$ an insecticide, BHC

In considering the possible reactivity of the halogenoalkanes, we should first look at the strength of the bonds and we shall use the bond energies:

$$CH_3Cl \longrightarrow CH_3 \cdot + Cl \cdot \qquad E(C-Cl) = 351\,kJ\,mol^{-1}$$

$$CH_3Br \longrightarrow CH_3 \cdot + Br \cdot \qquad E(C-Br) = 293\,kJ\,mol^{-1}$$

$$CH_3I \longrightarrow CH_3 \cdot + I \cdot \qquad E(C-I) = 234\,kJ\,mol^{-1}$$

so

$$E(C-Cl) > E(C-Br) > E(C-I)$$

If we change the molecular structure of the alkyl group, we find that the bond energy also changes:

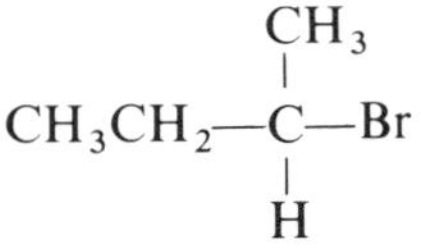

$$CH_3CH_2-\underset{\underset{H}{\vert}}{\overset{\overset{CH_3}{\vert}}{C}}-Br \qquad E(C-Br) = 284\,kJ\,mol^{-1}$$

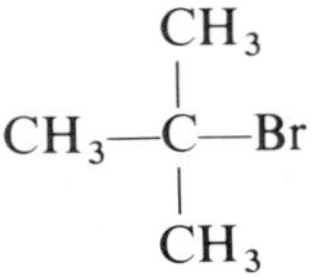

$$CH_3-\underset{\underset{CH_3}{\vert}}{\overset{\overset{CH_3}{\vert}}{C}}-Br \qquad E(C-Br) = 263\,kJ\,mol^{-1}$$

Secondly, we can look at the dipole moments of some molecules containing halogen atoms

	Dipole moment/D
1-chlorobutane	2.16
1-bromobutane	1.93
1-iodobutane	1.88

The existence of dipole moments in molecules has led to the suggestion that bonds can be polar. Dipole moments can only be a guide to polarities, since a dipole moment must be the sum of all the bond polarities, but we can suggest that

$$C^{\delta+}-Cl^{\delta-} > C^{\delta+}-Br^{\delta-} > C^{\delta+}-I^{\delta-}$$

On the basis of these data and your knowledge of halogen chemistry, what reagents might attack halogenoalkanes?

The characteristic infra-red stretching wavenumbers are affected by the relative atomic masses of the halogens.

	A_r (halogen)	Wavenumber/cm^{-1}
C—Cl	35.5	800–600
C—Br	80	600–500
C—I	127	500

The C—Br and C—I wavenumbers lie outside the usual range of infra-red spectrometers and this makes infra-red spectra less useful in the case of organic halogen compounds. The infra-red spectrum of 1-chlorobutane is shown in figure 9.8. The mass spectra of organic halogen compounds are interesting because they demonstrate the isotopic composition of chlorine and bromine. The ratio of ^{35}Cl to ^{37}Cl is about 3 to 1, while that of ^{79}Br to ^{81}Br is about 1 to 1.

The mass spectrum of 1-chlorobutane is shown in figure 9.9.

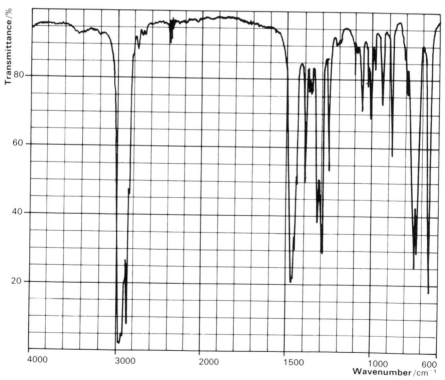

Figure 9.8
The infra-red spectrum of 1-chlorobutane.

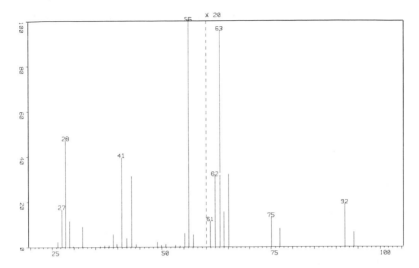

Figure 9.9
A reading from a mass spectrometer giving the mass spectrum of 1-chlorobutane.
Relative abundance is shown as a percentage on the vertical axis. It is multiplied by
20 for masses (shown on the horizontal axis) above 60.

Figures for the abundances of the principal fragments, relative to the
abundance of the fragment of mass 56.1, are given in the following table.

Mass	Relative abundance %	Mass	Relative abundance %
26.3	2.41	56.1	100.00
27.2	16.90	57.1	5.58
28.1	47.25	62.0	1.58
29.0	11.70	63.0	4.79
32.0	9.20	64.1	0.79
39.0	5.83	65.1	1.62
40.9	39.80	75.1	0.67
42.0	3.96	77.0	0.42
43.1	31.43	92.0	0.96
48.9	2.37	94.0	0.33
55.1	6.20		

Experiments with halogenoalkanes

In these experiments, we shall concentrate on reactions that are likely to occur with the halogen atom, remembering that ionic reagents are likely to be favoured. We shall compare the three halogenoalkanes

$$CH_3{-}CH_2{-}CH_2{-}CH_2Cl \qquad CH_3{-}CH_2{-}CH_2{-}CH_2Br \qquad CH_3{-}CH_2{-}CH_2{-}CH_2I$$

1-chlorobutane   1-bromobutane   1-iodobutane

and also look at the influence of the structures of the alkyl group.

The structures are known as primary, secondary, or tertiary on the basis of the number of alkyl groups joined to the carbon atom which is bonded to the halogen atom in the compound:

$$CH_3{-}CH_2{-}CH_2{-}CH_2Cl \qquad CH_3{-}CH_2{-}\underset{\underset{Cl}{|}}{CH}{-}CH_3 \qquad CH_3{-}\underset{\underset{Cl}{|}}{\overset{\overset{CH_3}{|}}{C}}{-}CH_3$$

1-chlorobutane   2-chlorobutane   2-chloro-2-methylpropane

(primary)   (secondary)   (tertiary)

EXPERIMENT 9.3
An investigation of the reactions of the halogenoalkanes

Procedure

Except where otherwise stated, use 2-chloro-2-methylpropane for the experiments, as it is much the cheapest of the halogenoalkanes you will be using.

Wear safety glasses throughout.

1 *Combustion* Keep the halogenoalkane in a test-tube well away from any flame. Dip a combustion spoon into it. Set fire to the halogenoalkane on the combustion spoon and note how readily or otherwise it burns.

2 *Reaction with aqueous alkali* To 1 cm^2 of 20% potassium hydroxide in ethanol (*TAKE CARE*), add an equal volume of water followed by 0.5 cm^3 of 2-chloro-2-methylpropane. Shake the tube from side to side for a minute. To test for chloride ions add an equal volume of 2M nitric acid to neutralize the potassium hydroxide (test with indicator paper to ensure that your solution is acidic) and then add a few drops of 0.02M silver nitrate. If chloride *ions* are present a white precipitate of silver chloride will appear.

What new organic compound has been formed?

3 *A comparison of halogenoalkanes*
a Arrange three test-tubes in a row and add 3 drops of halogenoalkane in the sequence 1-chlorobutane, 1-bromobutane, 1-iodobutane.

Now add $2\,cm^3$ of ethanol to each test-tube to act as a solvent. In each of three test-tubes, heat $2\,cm^3$ of 0.02M silver nitrate to near boiling and then, as quickly as possible, add $2\,cm^3$ to each halogenoalkane, in sequence, starting with 1-chlorobutane. Note the order in which precipitates appear and try to relate the reactivity of the halogenoalkanes to their dipole moments and to the halogen–carbon bond energies.

Which factor appears to be the more important when considering the rates of reaction?

b Repeat the experiment in a further three test-tubes using the halogenoalkanes 1-chlorobutane (primary), 2-chlorobutane (secondary), and 2-chloro-2-methylpropane (tertiary).

Comment on your results as you did for part **a**.

4 *Reaction with alcoholic alkali* To $2\,cm^3$ of 20% potassium hydroxide in ethanol (*TAKE CARE*) add $0.5\,cm^3$ of 2-chloro-2-methylpropane. Push a loose plug of ceramic fibres into the mixture and arrange the test-tube for collection of gas (figure 9.10). Heat gently and collect 2 to 3 test-tubes of gas. Test the gas for flammability; also test it with 2% bromine in 1,1,1-trichloroethane.

What new organic compound has been formed?

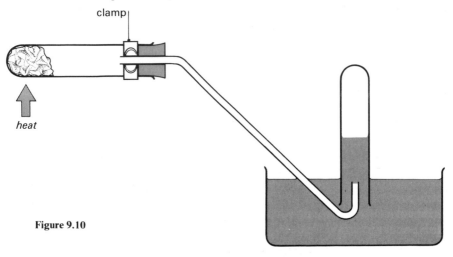

clamp

heat

Figure 9.10

An interpretation of the halogenoalkane experiment

We can write an equation for the reaction in part **2** of the experiment, between 2-chloro-2-methylpropane and potassium hydroxide in aqueous solution.

$$(CH_3)_3C—Cl + K^+OH^- \longrightarrow (CH_3)_3C—OH + K^+Cl^-$$

This reaction can tell us a great deal about organic halogen reactions in general. Firstly, we can see that a hydroxide ion has been exchanged for a chloride ion. Because of this, the reaction is known as a *substitution reaction*. Secondly, we need to consider the process by which the reaction occurred. We need to consider what species leaves the organic molecule, the *leaving group*, and what species attacks, the *attacking group*.

1 During the reaction a chloride ion has left the molecule. Being a chloride *ion* it will have taken with it the electron pair that formed the covalent bond

$$(CH_3)_3C \text{ :} \overset{..}{\underset{..}{Cl}} \text{:} \longrightarrow (CH_3)_3C^+ + \left[\text{:} \overset{..}{\underset{..}{Cl}} \text{:} \right]^- \text{ (charged)}$$

You should contrast this with the process that would produce a free radical

$$(CH_3)_3C \text{ :} \overset{..}{\underset{..}{Cl}} \text{:} \longrightarrow (CH_3)_3C\cdot + \cdot \overset{..}{\underset{..}{Cl}} \text{:} \text{ (uncharged)}$$

2 Since the chlorine atom takes away the bonding electrons, the group that attacks the tertiary chlorobutane molecule will need to have an unshared pair of electrons available for bonding to the carbon atoms, for example:

:O—H⁻

3 Because of the polarity of the carbon atom, $C^{\delta+}$, it will be an advantage for the attacking group to be negatively charged. *Attacking groups with an unshared pair of electrons available for forming a new covalent bond are known as nucleophiles.* In addition, they are often negatively charged. 'Nucleus' is Latin (meaning 'little nut') and the suffix '-phile' is derived from ancient Greek and means 'loving'; so 'nucleophile' means 'nucleus-loving'.

So far we have used the idea of bond polarity to interpret the process by which a halogenoalkane might react. Is polarity also a guide to the relative ease of reaction? Look at your results for part **3** of experiment 9.3, on the reaction between silver nitrate and six different halogenoalkanes. What are the leaving groups? What are the possible attacking groups? Can this be described as a 'nucleophilic substitution reaction'? If so, it follows that the discussion about the reaction between potassium hydroxide and 2-chloro-2-methylpropane is also applicable to these reactions.

Now compare bond polarities (page 275) with the relative ease of formation of the silver halide precipitates. Finally, compare bond energies (page 275) with the relative ease of formation of the precipitates.

> Which factor, bond polarity or bond strength, is the best guide to ease of reaction in this particular case?

The reaction in part **4** of experiment 9.3 is of a different type. The gaseous product was an unsaturated hydrocarbon, which means that both halogen and hydrogen have been lost from the 2-chloro-2-methylpropane molecule.

$$
\begin{array}{ccc}
\quad\ \ \overset{\displaystyle CH_3}{\underset{}{|}} & & \overset{\displaystyle CH_3}{\underset{}{|}}\\[2pt]
CH_3-\!\!\underset{\underset{Cl}{|}}{\overset{}{C}}\!\!-\!\!\underset{\underset{H}{|}}{\overset{}{C}}H_2 & \longrightarrow & CH_3-\!\!\overset{}{C}\!\!=\!\!CH_2 + HCl
\end{array}
$$

This is known as an *elimination reaction*. Chemists think that the hydroxide ion, a powerful base, extracts a proton from the halogenoalkane and a chloride ion separates from the molecule. Notice that elimination does not occur by the departure of hydrogen and chlorine as hydrogen chloride in the same step in the reaction process. The overall reaction can be written as:

$$
\begin{array}{ccc}
\quad\ \ \overset{\displaystyle CH_3}{\underset{}{|}} & & \overset{\displaystyle CH_3}{\underset{}{|}}\\[2pt]
CH_3-\!\!\underset{\underset{Cl}{|}}{\overset{}{C}}\!\!-\!\!CH_3 + K^+OH^- & \longrightarrow & CH_3-\!\!\overset{}{C}\!\!=\!\!CH_2 + K^+Cl^- + H_2O
\end{array}
$$

Reactions of the halogenoalkanes

1 *Substitution by a hydroxyl group*

$$CH_3-CH_2-CH_2-CH_2Br + K^+OH^- \longrightarrow$$
$$CH_3-CH_2-CH_2-CH_2OH + K^+Br^-$$

This equation can be written more briefly as

$$CH_3(CH_2)_3Br + OH^- \longrightarrow CH_3(CH_2)_3OH + Br^-$$

The hydroxide ion is a strong nucleophile, attacking the terminal carbon atom and substituting for the bromine atom. The reaction is, therefore, described as a nucleophilic substitution.

The reaction is not used to prepare the common alcohols because they are more readily and cheaply prepared by other reactions (see for example number **3** in the reactions of alkenes, at the end of section 9.4) but the reaction may be useful when chemists want to substitute a hydroxyl group into a complex compound.

2 *Substitution by an amine group*

$$CH_3(CH_2)_3Br + 2:NH_3 \longrightarrow CH_3(CH_2)_3NH_2 + NH_4^+Br^-$$
butylamine

In this reaction ammonia uses an unshared pair of electrons and therefore functions as a nucleophile. An alcoholic solution of ammonia is needed, and heating is carried out under pressure to give an adequate concentration of ammonia. Yields are not good and the reaction is complicated by the formation of secondary and tertiary alkyl amines, R_2NH and R_3N (see Topic 13), since the amines themselves are nucleophiles.

3 *Substitution by a nitrile group*

$$CH_3(CH_2)_3Br + Na^+CN^- \longrightarrow CH_3(CH_2)_3CN + Na^+Br^-$$
pentanenitrile

This is a valuable reaction for the synthesis of other compounds because the reaction increases the number of carbon atoms in the chain. Good yields of nitriles are obtained by refluxing with sodium cyanide in ethanol. The nucleophile is the cyanide ion, $:C{\equiv}N^-$, with the lone pair of electrons on the carbon atom forming the new bond.

4 *Elimination in the synthesis of alkenes*

$$\underset{\substack{\text{2-bromo-2-}\\\text{methylpropane}}}{(CH_3)_3CBr} + K^+OH^- \longrightarrow \underset{\text{methylpropene}}{\overset{\overset{\textstyle CH_3}{|}}{CH_3CH{=}CH_2}} + K^+Br^- + H_2O$$

This is a good method of introducing double bonds into complex molecules. The reagent, concentrated potassium hydroxide (a strong base), is the same as the one used to substitute for halogen (in reaction **1**). But because the solvent is changed from ethanol and water to ethanol alone, and because the reaction

but if we consider just the π bond we find it is weaker than a σ bond. By calculation from the data above, you should be able to confirm that the difference between the two bond energies is only $264\,kJ\,mol^{-1}$.

The infra-red spectra of alkenes are significantly different from those of alkanes, as might be expected from the differences in structure (figure 9.12). In the infra-red spectrum of oct-1-ene there is an absorption peak due to $C{=}C$ stretch at $1650\,cm^{-1}$ and another peak at $1825\,cm^{-1}$. There is also a peak due

to $C{-}H$ stretch in $=C\overset{\displaystyle H}{\underset{\displaystyle H}{\diagdown}}$ at $3100\,cm^{-1}$. The other peaks are mainly due to

CH_3 and CH_2 groups.

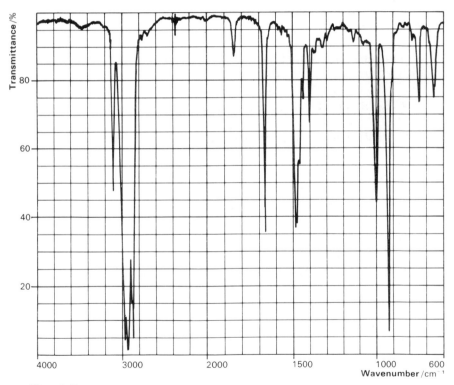

Figure 9.12
The infra-red spectrum of oct-1-ene.

The rules for naming unsaturated hydrocarbons are similar to those used for the corresponding saturated compounds, explained in section 9.1. The only additional problem which arises is that of locating the double bond. This is done by using the carbon atom of lower number, of the pair of carbon atoms connected by the double bond.

$CH_2{=}CH{-}CH_2{-}CH_3$ but-1-ene (not but-2-ene)

There is no need to do this for the first two members of the series $CH_2{=}CH_2$ ethene (also called ethylene), or for $CH_3{-}CH{=}CH_2$ propene (also called propylene).

With four carbon atoms and one double bond structural isomers are possible.

$CH_2{=}CH{-}CH_2{-}CH_3$ but-1-ene
$CH_3{-}CH{=}CH{-}CH_3$ but-2-ene

The same rule is used for compounds containing more than one double bond, for example

$CH_2{=}CH{-}CH{=}CH{-}CH_3$ penta-1,3-diene

(The 'di' indicates two double bonds; note also that 'a' is added to the hydrocarbon root when more than one double bond is present.)

Branched chain alkenes are dealt with as for alkanes, for example

$CH_2{=}\overset{|}{\underset{CH_3}{C}}{-}CH_2{-}CH_3$ is 2-methylbut-1-ene

Cycloalkenes follow similar rules to cycloalkanes.

or   is cyclohexene

Experiments with alkenes

For these experiments we shall repeat most of the reactions carried out with alkanes in section 9.2, but we shall be using unsaturated instead of saturated hydrocarbons.

EXPERIMENT 9.4
An investigation of the reactions of the alkenes

For this experiment you should use the alkenes cyclohexene and limonene. Handle cyclohexene with care: it is highly flammable.

cyclohexene
obtained from petroleum

limonene
extracted from oranges

If you do part **5** of the experiment first, you will be able to use your own sample of limonene.

Record your results in the form of a table, so that they can be compared with the results of the similar experiments with the alkanes.

Wear safety glasses throughout the experiment. Handle 1,1,1-trichloro-ethane in a fume cupboard.

Procedure

1 *Combustion* Keep the liquid alkene in a small test-tube well away from any flame. Dip a combustion spoon into the sample. Set fire to the alkene on the combustion spoon and note the luminosity and sootiness of the flame.

2 *Oxidation* To 0.5 cm³ of the alkene in a test-tube add a few drops of a mixture of equal volumes of 0.01M potassium manganate(VII) solution and 2M sulphuric acid. Shake the contents and try to tell from any colour change of the manganate(VII) if it oxidizes the alkene.

3 *Action of bromine* To 0.5 cm³ of the alkene in a test-tube add a few drops of 2% bromine dissolved in 1,1,1-trichloroethane. (*TAKE CARE.*)

What happens to the colour of the bromine?

4 *Action of sulphuric acid* Put 1–2 cm³ of concentrated sulphuric acid (*TAKE CARE*) in a test-tube held in a test-tube rack and add 0.5 cm³ of the alkene. Shake the test-tube *gently*.

Do the substances mix or are there two separate layers in the test-tube?

5 *Laboratory preparation: extraction of limonene from oranges* For the experiments on alkenes you can use either the aqueous mixture from the steam distillation or the 1,1,1-trichloroethane solution from the extraction, but to get results you may need to use more than the $0.5\,cm^3$ of alkene suggested for experiments **2** to **4**.

Put into a $250\,cm^3$ flask the finely ground or chopped outer rind of 2 oranges and $100\,cm^3$ of water. This is the minimum amount to use but the extraction can be scaled up to suit any size of container. Use only the outer, orange-coloured rind, which needs to be fresh. Arrange the flask for distillation (figure 9.13) and heat on a wire gauze. Collect about $50\,cm^3$ of distillate in a measuring cylinder. You should be able to see an oily layer of limonene on top of the water. (If you are stopping at this stage, use a dropping pipette to remove portions of the oily layer for experiments **2**, **3**, and **4** above.)

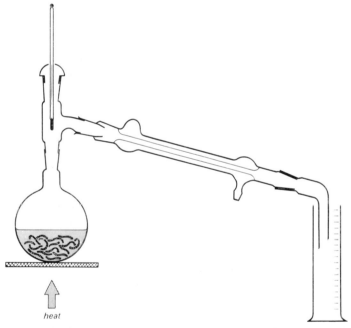

heat

Figure 9.13

Transfer the distillate to a separating funnel and add $20\,cm^3$ of 1,1,1-trichloroethane. Shake the funnel to mix the two layers to extract the limonene into the 1,1,1-trichloroethane. After shaking for one minute, allow the two layers to separate and run the lower 1,1,1-trichloroethane layer (density $1.32\,g\,cm^{-3}$) into a small conical flask, being careful to let none of the water layer escape. Add a few spatula measures of anhydrous sodium sulphate to dry the 1,1,1-trichloroethane solution. (If you are stopping at this stage, use 5 cm^3 of your 1,1,1-trichloroethane solution for each experiment **2**, **3**, and **4** above.) Meanwhile set up a clean dry apparatus for distillation. Filter the 1,1,1-trichloroethane solution through a fluted filter paper (see figure 9.14) into the distillation flask and distil off the 1,1,1-trichloroethane ($T_b = 74°C$). Stop heating when about 2 cm^3 of liquid remain in the flask and the rate of boiling suddenly slows down. If you do not stop now you will vaporize the product and have to wait for the flask to cool before it can be seen. The limonene ($T_b = 176°C$) will remain in the flask. The yield from two oranges should be about 1 cm^3.

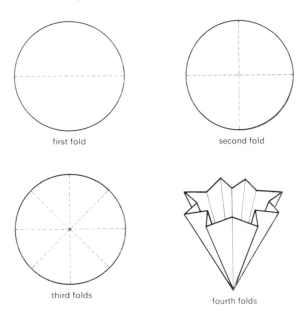

first fold

second fold

third folds

fourth folds

Figure 9.14
'Fluting' the filter paper. Open the filter paper out after each fold so that the first, second, and third folds are all in the same direction. The fourth set of folds must be in the opposite direction.

It is not usual to be able to extract a natural product in a pure state from natural sources so readily.

An interpretation of the experiments with alkenes

It should have become obvious to you that the alkenes are readily reactive to substances such as concentrated sulphuric acid and bromine, whereas the alkanes are unreactive. Furthermore, it can be shown that the reactions are giving only a single product. Since the reagents are adding to the C=C double bond and not removing any atoms from the alkene, the reactions are known as *addition reactions:*

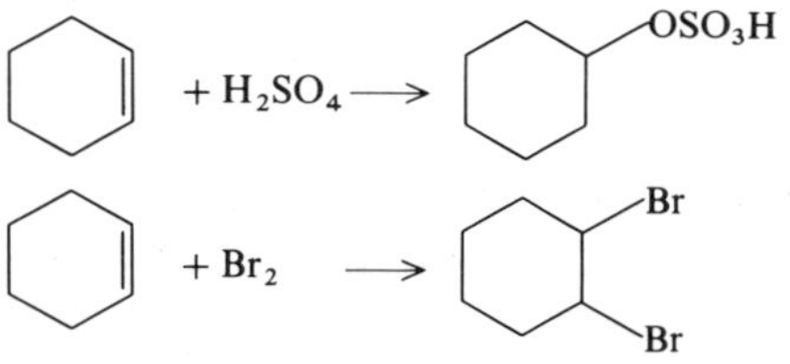

By what process do these reactions occur? From the study of a large number of alkene reactions chemists have proposed a mechanism that is consistent with all the evidence available. The π bond contributes an electron pair to the formation of a new bond with the positively polarized attacking group.

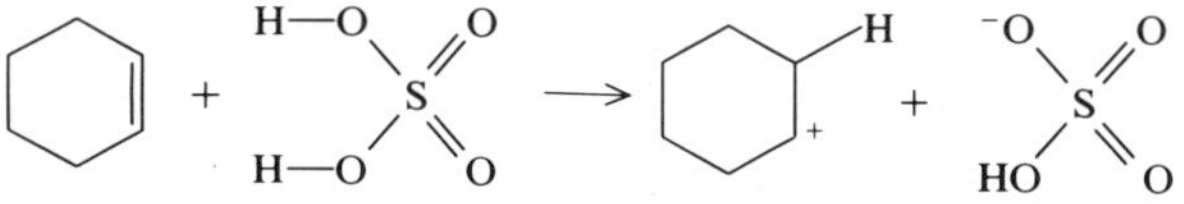

In this and similar reactions the attacking group is called an *electrophile* (electron-seeking). Electrophiles are commonly acidic compounds, as in the above reaction in which sulphuric acid is reacting as an electrophile. The sulphuric acid donates a proton, H^+, to the cyclohexene molecule, which provides an electron pair to form the new bond to the proton.

This is the common pattern of electrophilic reactions. The electrophile is an electron-deficient compound that can form a new covalent bond, using an electron pair provided by the carbon compound. The commonest *electrophilic reagent* is the proton, H^+.

The reaction ends with the addition of a hydrogensulphate ion to the cyclohexyl ion, because a positively charged carbon atom in the cyclohexyl ring (called a *carbocation*) is still a reactive species.

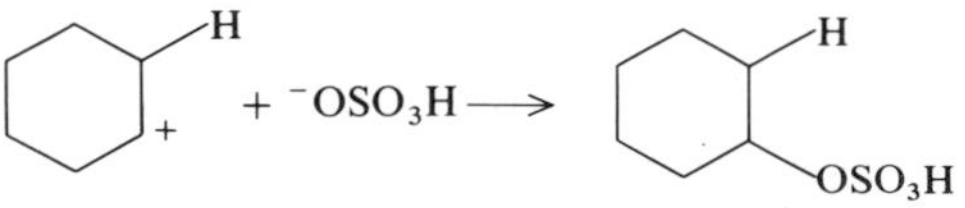

cyclohexyl
hydrogensulphate

This type of reaction is described as an electrophilic addition.

The reaction with bromine may not follow the same course because the bromine molecule is not polar and does not have a proton to donate to the double bond. Read the passage quoted below and try to decide whether the results can be explained in terms of the course described for the electrophilic addition of sulphuric acid or whether we shall have to propose a different process for the addition of bromine.

 ' A $1 dm^3$ pressure bottle was filled two-thirds full with a saturated salt solution, and sufficient halogen was added to saturate it. The bottle was closed with a cap containing a bicycle valve, and a moderate pressure (4 or 5 atmospheres) of ethene was added from a cylinder. The bottle was well shaken for one minute or until the pressure had become practically atmospheric, and more ethene was added. When the solution had become colourless, more halogen and ethene were introduced. The process was continued until a sufficient amount of oil had been accumulated. This was separated from the aqueous solution, washed with water, dried with calcium chloride, and examined by determination of density, refractive index, or boiling point. In each case a mixture was obtained, and a partial separation was made by fractional distillation.

 ' From ethene, bromine, and sodium chloride solution a mixture of 1,2-dibromoethane and 1-bromo-2-chloroethane was obtained. The product mixture, an oil, contained about 46 per cent of C_2H_4ClBr as estimated from the refractive index, 1.51 (the value for $C_2H_4Br_2$ in the literature is 1.53; for $C_2H_4Cl_2$, 1.44).

 ' 1-bromo-2-nitratoethane was obtained from ethene, bromine, and sodium nitrate solution. Before distillation the product mixture was washed with sodium hydrogen carbonate solution to remove any trace of nitric acid. The oil began to boil at 132 °C ($C_2H_4Br_2$) but a portion boiled at 163–5 °C (the boiling point of $C_2H_4BrNO_3$ given in the literature is 164 °C). In the distillation the last trace exploded with evolution of brown fumes of nitrogen dioxide, recognized also by their odour.'

These results were obtained by A. W. Francis, and published in the *Journal of the American Chemical Society* in 1925.

An equation for one reaction carried out by Francis could be written as:

$$CH_2{=}CH_2 + Br_2 + NO_3^- \longrightarrow CH_2Br{-}CH_2ONO_2 + Br^-$$

Can you see how this product might have come about?

Consider first the nitrato ($-ONO_2$) group, which is derived from the nitrate (NO_3^-) ion.

> Would you describe the nitrate ion as a free radical, an electron-deficient or an electron-rich species?

> So would the nitrate ion be most likely to attack a carbon grouping of the free radical type, a π bond, or a carbocation?

> Now consider the bromo ($-Br$) group, which is derived from the bromine (Br_2) molecule.

> Can a bromine–bromine bond break to give free radicals, electron-deficient and electron-rich species?

> Which of these would react with a π bond to give a molecule with a charge that the nitrate ion might attack?

> Finally, does the process you propose give a bromide (Br^-) ion as one of the products?

Reactions of the alkenes

1 *Addition of halogens*

$$CH_3-CH{=}CH_2 + Br_2 \longrightarrow CH_3-CHBr-CH_2Br$$

1,2-dibromopropane

This is an electrophilic addition with the Br—Br molecule being polarized in part by the π bond. The reaction needs to be carried out without heat and in the absence of sunlight, to avoid free radical substitution (see section 9.2). It is an important reaction for the preparation of dihalogenoalkanes.

2 *Addition of hydrogen halides* In normal laboratory conditions the hydrogen halide acts as an electrophile, forming the intermediate $CH_3-\overset{+}{C}H-CH_3$ with the positive charge on the —CH— group.

$$CH_3-CH{=}CH_2 + HBr \longrightarrow CH_3-CHBr-CH_3$$

2-bromopropane

In sunlight or with a suitable catalyst the alternative product, 1-bromopropane, is obtained.

3 *Addition of sulphuric acid* This is an important reaction because the product, an alkyl hydrogensulphate, will take part in further reactions producing alcohols as shown in the following reaction scheme. These reactions are carried out in industry.

$$CH_2{=}CH_2 \xrightarrow{H_2SO_4} CH_3{-}CH_2{-}OSO_3H \xrightarrow{H_2O} CH_3{-}CH_2{-}OH$$
ethanol

The addition of sulphuric acid follows the same pattern as the electrophilic addition of hydrogen halides.

4 *Formation of diols by potassium manganate*(VII) Reaction with cold acidified potassium manganate(VII) produces compounds known as *diols*. The complete balanced equation for this reaction is complicated. As the interest is chiefly centred on the organic compounds, we can write a simplified version in this way.

$$CH_2{=}CH_2 \xrightarrow{KMnO_4} CH_2OH{-}CH_2OH$$
ethane-1,2-diol

The reaction involves a change in the oxidation number of each carbon atom from -2 to -1. An alternative name for the product is glycol. It is commonly used in anti-freeze for car radiators and is manufactured by a more efficient method than manganate(VII) oxidation.

5 *Reduction with hydrogen* Reduction by hydrogen requires the use of a metal catalyst such as nickel. The metal has to be finely divided, and in the case of nickel the catalyst is made by treating a special nickel–aluminium alloy with sodium hydroxide. This dissolves away the aluminium and leaves the nickel (known as Raney nickel) in a very finely divided state.

$$CH_2{=}CH_2 + H_2 \xrightarrow[\text{catalyst}]{Ni} CH_3{-}CH_3$$

The reaction is useful for preparing alkanes or for saturating some of the double bonds in natural oils. In this way, liquid fats can be converted to solids for use in margarine (although saturated fats are considered to be a contributor to heart complaints and unsaturated fats safer to eat).

6 *Polymerization* Under conditions of high temperature and pressure, and in the presence of various catalysts, many alkenes undergo *polymerization*. A large number of molecules of the alkene (the *monomer*) join together to make molecules of a polyalkene (the *polymer*). The molecules of the polymer have the same empirical formula as those of the monomer. Ethene, for example, polymerizes to form poly(ethene).

$$n\,CH_2{=}CH_2 \longrightarrow {+}CH_2{-}CH_2{+}_n$$

Several of the products of this type of reaction are commercially important polymers, and will be considered further in Topic 17.

BACKGROUND READING
Octane number of petrol hydrocarbons

When it is first obtained from the Earth, crude oil is a complex mixture of hydrocarbons with sulphur compounds and inorganic impurities. The hydrocarbons may contain one to more than fifty carbon atoms, and are mostly alkanes (with straight or branched chains), together with naphthenes and arenes. Petroleum is separated into fractions by distillation, for example, gasoline, naphtha, kerosine, gas oil, and Diesel oil.

In general, the percentage of motor gasoline or petrol in crude oils is not enough to meet the heavy demands for motor car use. So it is necessary to devise ways whereby a larger proportion of the hydrocarbons in crude oil can be made use of as petrol. The value of hydrocarbons for use in petrol can be judged from their 'octane number'. Heptane is given an octane number of 0 and 2,2,4-trimethylpentane (iso-octane) is given an octane number of 100. The higher the number, the less the tendency to pre-ignite in a car engine – that is, the less the tendency to explode under compression before the spark is passed. A second explosion when the spark is passed results in the two shock waves producing a characteristic 'knocking' in the engine.

Four processes of importance for producing petrol-grade hydrocarbons are catalytic cracking, catalytic reforming, alkylation, and isomerization.

Catalytic cracking One method of obtaining more petrol is to heat the larger hydrocarbon molecules so that they break down. In early years the process of thermal cracking was used, although much of the petroleum was broken down too extensively. In the 1930s, the higher compression in petrol engines called for fuels with a higher octane rating. The value for the products of thermal cracking was only 70–80. Fortunately, it had been discovered that the cracking of hydrocarbons in the presence of a catalyst (catalytic cracking) gave a petrol containing more branched hydrocarbons and an octane rating of 90–95.

The first catalytic cracking unit was built in 1936 in the USA, at New Jersey. It contained a fixed bed of catalyst pellets composed of acid-treated clays. From a knowledge of the chemical composition of clays, various synthetic silica–alumina catalysts were also developed and these are still widely used. More recently, crystalline aluminosilicates, known as zeolites or molecular sieves, have also come into use as cracking catalysts.

Catalytic reforming. This is now one of the most important processes for the production of motor gasolines. Adding a metallic component to a cracking catalyst gives petrol with an even higher octane number. Platinum is used exclusively as this component and highly purified alumina is used in place of silica–alumina. The process is known as catalytic reforming, but 'platforming' and other commercial names are often used. The improvement in octane number is due largely to the higher percentage of arenes in the product. The process is therefore also a source of arenes for the chemical industry. Some of the chemical reactions which are carried out at the same time by reforming catalysts are:

dehydrogenation of cyclohexanes to arenes;
dehydrocyclization of alkanes and alkenes to arenes;
isomerization of unbranched chain to branched chain alkanes;
hydrocracking to hydrocarbons of lower relative molecular mass.

In a reforming catalyst, the platinum is highly dispersed over the alumina, perhaps as platinum atoms or small groups of atoms. Both the platinum and the alumina play a catalytic role.

Alkylation. Another means of obtaining high octane blending stocks is to join some of the smaller molecules in the right way, that is, using C_3—C_4 hydrocarbons. The process of alkylation involves the reaction of a branched chain alkane (for example 2-methylpropane) and an alkene (for example propene or butene). The catalysts are or contain acids; sulphuric acid, hydrofluoric acid, and phosphoric acid are used.

Isomerization. As alkanes with branched chains have a higher octane number than those with straight chains, a process for converting straight C_5 or C_6 chains to branched chains has been developed. The catalyst used is a specially prepared platinum material kept in an active state by adding an activator to the reactants.

Octane number and molecular structure

The relationship of octane number to molecular structure can be seen in the tables of the C_7 alkanes, the cyclic compounds, and the C_7 alkenes.

C_7 alkanes	Octane number	C_7 alkanes	Octane number
Heptane	0	2,4-dimethylpentane	77
2-methylhexane	41	3,3-dimethylpentane	95
3-methylhexane	56	3-ethylpentane	64
2,2-dimethylpentane	89	2,2,3-trimethylbutane	113
2,3-dimethylpentane	87		

You can see that the more branches to the carbon chain the higher the octane number and hence the value of isomerization:

$$CH_3-CH_2-CH_2-CH_2-CH_2-CH_2-CH_3 \xrightarrow{\text{isomerization}} CH_3-CH_2-CH_2-\overset{\displaystyle CH_3}{\underset{\displaystyle CH_3}{C}}-CH_3$$

octane number 0

octane number 89

Cyclic compounds	Octane number
(Hexane)	(26)
Cyclohexane	77
Methylcyclohexane	104
Benzene	108
Methylbenzene	124

The conversion of alkanes to cycloalkanes and the dehydrogenation of cycloalkanes by the process of catalytic reforming also enhance the octane number of the petrol fraction from crude oil:

$$CH_3-CH_2-CH_2-CH_2-CH_2-CH_3 \xrightarrow[\text{reforming}]{\text{catalytic}} \bigcirc + H_2$$

octane number 26

octane number 77

$$\bigcirc \xrightarrow[\text{reforming}]{\text{catalytic}} \bigcirc\!\!\!\!\bigcirc + 3H_2$$

octane number 108

C_7 alkenes	Octane number	C_7 alkenes	Octane number
Hept-1-ene	68	4,4-dimethylpent-1-ene	144
5-methylhex-1-ene	96	2,3-dimethylpent-2-ene	165
2-methylhex-2-ene	129	2,4-dimethylpent-2-ene	135
2,4-dimethylpent-1-ene	142	2,2,3-trimethylbut-1-ene	145

If you compare this table of C_7 alkenes with the previous one of C_7 alkanes, you can see that the formation of an unsaturated compound enhances the octane number of a hydrocarbon.

$$CH_3-CH_2-CH_2-\overset{\overset{\displaystyle CH_3}{|}}{\underset{\underset{\displaystyle CH_3}{|}}{C}}-CH_3 \xrightarrow[\text{reforming}]{\text{catalytic}} CH_2=CH-CH_2-\overset{\overset{\displaystyle CH_3}{|}}{\underset{\underset{\displaystyle CH_3}{|}}{C}}-CH_3 + H_2$$

octane number 89   octane number 144

Why do these changes in molecular structure enhance octane numbers? The answer lies in the process of combustion. The conditions of temperature and pressure in a car engine result in the production of free radicals. The more reactive the free radicals, the greater the chance of an uncontrolled chain reaction such as pre-ignition explosion or knocking in the engine.

$$CH_3-CH_2-CH_2-CH_2-CH_2-CH_3 \longrightarrow 2CH_3-CH_2-CH_2\cdot$$

octane number 26   reactive radicals

$$CH_3-\overset{\overset{\displaystyle CH_3}{|}}{\underset{\underset{\displaystyle CH_3}{|}}{C}}-CH_2-\overset{\overset{\displaystyle CH_3}{|}}{CH}-CH_3 \longrightarrow CH_3-\overset{\overset{\displaystyle CH_3}{|}}{\underset{\underset{\displaystyle CH_3}{|}}{C}}-CH_2\cdot + \cdot\overset{\overset{\displaystyle CH_3}{|}}{CH}-CH_3$$

octane number 100   less reactive radicals

The function of the petrol additive, tetraethyl lead, is to help to control the free radical chain reaction. When free radicals react with tetraethyl lead the chain is terminated because the final product is an unreactive lead atom:

$$(CH_3CH_2)_4Pb + 4\cdot\overset{\overset{\displaystyle CH_3}{|}}{CH}-CH_3 \longrightarrow 4CH_3-CH_2-\overset{\overset{\displaystyle CH_3}{|}}{CH}-CH_3 + \cdot\overset{\displaystyle ..}{Pb}\cdot$$

If the hydrocarbons of low octane number which cause knocking were not present in petrol there would be no need for the addition of tetraethyl lead.

9.5
BENZENE AND SOME SUBSTITUTED BENZENE COMPOUNDS

The structure of benzene, C_6H_6, provided chemists with a major problem. The principal difficulty was the absence of isomers of monosubstituted derivatives of benzene, such as chlorobenzene, C_6H_5Cl. An acceptable structure must therefore be one in which all six hydrogen atoms would occupy equivalent positions.

A major step towards the solution to the problem was taken by Kekulé, then Professor of Chemistry at Ghent in Belgium, in 1865. He later described how he came to propose the structure illustrated below.

'I turned my chair to the fire and dozed. Again the atoms were gambolling before my eyes. This time the smaller groups kept modestly in the background. My mental eye, rendered more acute by repeated visions of this kind, could now distinguish larger structures, of manifold conformation; long rows, sometimes more closely fitted together; all twining and twisting in snake-like motion. But look! What was that? One of the snakes had seized hold of its own tail, and the form whirled mockingly before my eyes. As if by a flash of lightning I awoke.'

Translation from FINLAY, ALEXANDER (1937) 100 years of chemistry. *Duckworth.*

Arthur Koestler, in his book *The act of creation* (Hutchinson, 1964), describes this as probably the most important dream in history since the interpretation by Joseph of Pharoah's dream of seven fat and seven lean cows (Genesis, chapter 40). 'The serpent biting its own tail', he writes, 'gave Kekulé the clue to a discovery which has been called "the most brilliant piece of prediction to be found in the whole range of organic chemistry", and which, in fact, is one of the cornerstones of modern science.' It was the first suggestion that carbon atoms in molecules formed not only chains, but also rings, like the snake swallowing its tail.

Figure 9.15
These stamps were issued to commemorate the centenary of Kekulé's proposal of a ring structure for benzene.

The modern evidence for the symmetry of the benzene ring is based on X-ray diffraction studies. The unusual nature of the bonding is seen from a comparison of the bond lengths of benzene with those of cyclohexene.

carbon–carbon single bond in cyclohexane   0.15 nm
carbon–carbon double bond in cyclohexene   0.13 nm
carbon–carbon bonds of benzene   0.14 nm

The bonding in benzene cannot therefore be described as three double bonds plus three single bonds, but must be considered as a delocalized electron cloud spread out over the whole ring, as in figure 9.16.

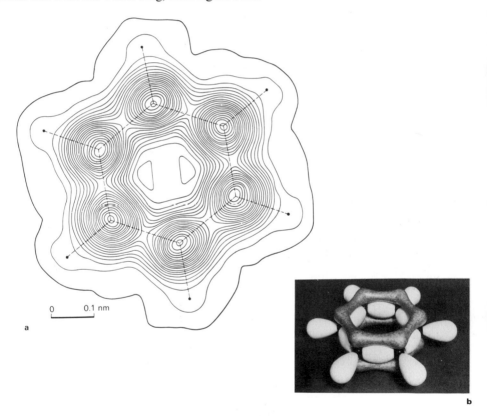

Figure 9.16
a Electron density map of benzene at $-3\,°C$. Contours are at 0.25 electron per $10^{-30}\,m^3$.
b A PEEL model showing the delocalized electron cloud in benzene.
a: *COX, E. G., CRUICKSHANK, D. W. J., and SMITH, J. A. S., 'Crystal structure'* Proc. Roy. Soc.
A. 247, 1958.
b: *Model, Griffin & George Ltd; photograph, University of Bristol, Faculty of Arts Photographic Unit.*

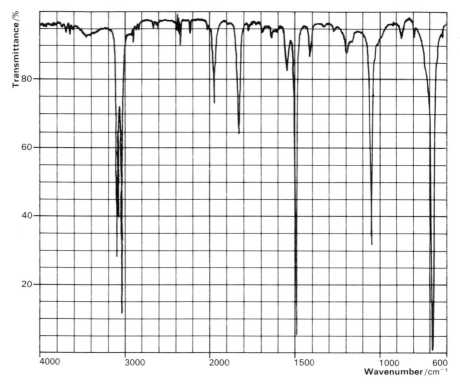

Figure 9.17
The infra-red spectrum of benzene.

When drawing a structure to indicate the molecule of benzene certain difficulties arise; a single line is normally used to represent two electrons, and two lines to represent four electrons. As neither of these is appropriate for the carbon–carbon bonds in benzene, this representation is often used:

Thermochemical data

The influence of the structure of benzene on its reactions can be looked at by considering the enthalpy change which takes place when hydrogen is added.

You have already seen that cyclohexene reacts with hydrogen to form cyclohexane.

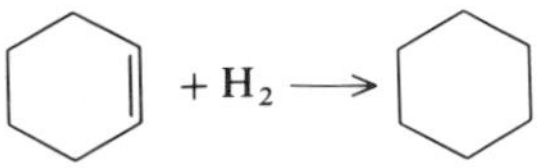

$$\text{(benzene ring structure)} + H_2 \longrightarrow \text{(cyclohexane)} \qquad \Delta H^{\ominus} = -120 \, \text{kJ mol}^{-1}$$

Use these data to calculate the enthalpy change of hydrogenation for a molecule with the Kekulé structure

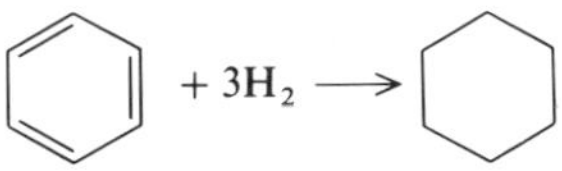

$$\text{(Kekulé structure)} + 3H_2 \longrightarrow \text{(cyclohexane)} \qquad \Delta H^{\ominus} = ?$$

You can compare your result with the known value for benzene.

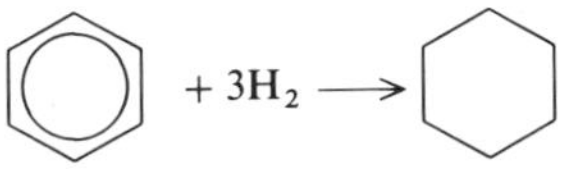

$$\text{(benzene)} + 3H_2 \longrightarrow \text{(cyclohexane)} \qquad \Delta H^{\ominus} = -208 \, \text{kJ mol}^{-1}$$

On the basis of this result, it is reasonable to deduce that the benzene ring is less likely to take part in addition reactions than other unsaturated compounds would be.

The infra-red spectrum of benzene is shown in figure 9.17 opposite.

The naming of arenes

Arenes were originally called the aromatic hydrocarbons. Two examples are:

benzene (C_6H_6) 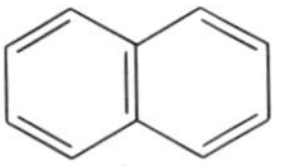   naphthalene $(C_{10}H_8)$

The group $C_6H_5\!-\!$, derived from benzene, is known as the phenyl group. Many substitution products, when one substituent only is involved, are commonly known by non-systematic names, for example

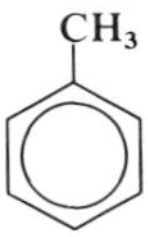

CH_3

methylbenzene (not phenylmethane); also known as toluene

When drawing such structures the convention is that where a group is attached to the benzene ring a hydrogen atom has been removed.

In the next section you will be doing some experiments to compare the reactions of some arenes.

Experiments with arenes

Benzene has been shown to be toxic and mildly carcinogenic. We shall therefore need to do our experiments with various derivatives of benzene such as methylbenzene and methoxybenzene.

methylbenzene (toluene)   methoxybenzene (anisole)

The hazards involved in experimenting with benzene have not been recognized for long and in older books you may find suggestions for experiments involving benzene itself. Such experiments should be avoided.

Although safer than benzene itself, the substitutes used in this experiment are flammable and have a harmful vapour. Take due care when handling them.

Record your results in such a way that you can compare them with those of the experiments done with alkanes (9.2) and with alkenes (9.4). The experiments with methylbenzene will enable you to compare the reactivity of arenes with that of alkanes and alkenes. In the experiments with methoxybenzene you will be able to consider the nature of the reactions of arenes. The methyl group and the methoxy group are unreactive in the conditions of the experiments, so any reactions you observe are likely to be reactions of the benzene ring.

EXPERIMENT 9.5a
An investigation of the reactions of the arenes
Wear safety glasses throughout the experiment.

Procedure for methylbenzene

1 *Combustion* Keep a sample of liquid methylbenzene in a test-tube well away from any flame. Dip a combustion spoon into the sample. Set fire to the methylbenzene on the combustion spoon and note the luminosity and sootiness of the flame.

2 *Oxidation* To $0.5\,cm^3$ of methylbenzene in a test-tube add a few drops of a mixture of equal volumes of 0.01M potassium manganate(VII) solution and 2M sulphuric acid. Shake the contents of the tube and try to tell from any colour change of the manganate(VII) whether it oxidizes the methylbenzene.

3 *Action of bromine.* To 1–2 cm^3 of methylbenzene in a test-tube add a few drops of 2% bromine dissolved in 1,1,1-trichloroethane. (*TAKE CARE.*)

What happens to the colour of the bromine?

4 *Action of sulphuric acid* Place 1–2 cm^3 of concentrated sulphuric acid (*TAKE CARE*) in a test-tube held in a test-tube rack and add 0.5 cm^3 of the methylbenzene. Shake the test-tube *gently*.

Do the substances mix or are there two separate layers in the test-tube?

Compare your results with those obtained with alkanes and with alkenes. You should be able to see that the benzene ring is comparable to the alkanes in stability. Also, it is remarkably resistant to the reagents that readily took part in addition reactions with alkenes.

Procedure for methoxybenzene

Methoxybenzene has a benzene ring that is fairly reactive, so by using appropriate reagents you should be able to observe some typical reactions of the benzene ring.

1 *Bromination* To 0.5 cm^3 of methoxybenzene in a test-tube add 1 cm^3 of 2% bromine in 1,1,1-trichloroethane.

What happens to the colour of the bromine? What are the fumes that are evolved (test them with ammonia)? Did alkenes give off fumes in this reaction? Has an addition reaction occurred?

2 *Sulphonation* To 0.5 cm^3 of methoxybenzene in a test-tube add 1 cm^3 concentrated sulphuric acid (*TAKE CARE*). Shake the test-tube *gently* to mix the contents (does the tube get hot?) then *cautiously* add 4 cm^3 water.

Is there a product which is soluble in water?

3 *Friedel–Crafts reaction* To 1 cm^3 of methoxybenzene in a test-tube add a small spatula measure of *anhydrous* aluminium chloride (*TAKE CARE*) followed by 1 cm^3 of 2-chloro-2-methylpropane. If necessary, warm the mixture in a beaker of hot water.

What are the fumes that are evolved (test with moist indicator paper)?

4 *Nitration* To 1 cm^3 of water add 1 cm^3 of concentrated nitric acid (*TAKE CARE*) followed by a few drops of methoxybenzene. Warm in a water bath and observe the formation of coloured products.

The difference between alkene reactions and benzene ring reactions can be seen most clearly in the bromination reaction. An alkene such as cyclohexene undergoes an addition reaction with bromine

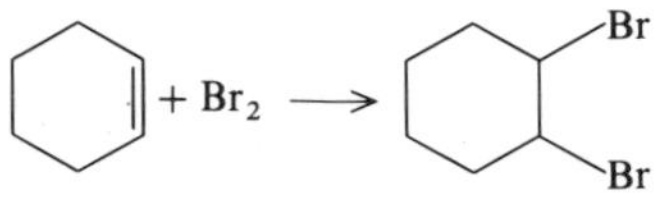

But when a benzene ring reacts, hydrogen bromide is produced and this means that a hydrogen atom has been displaced from the benzene ring:

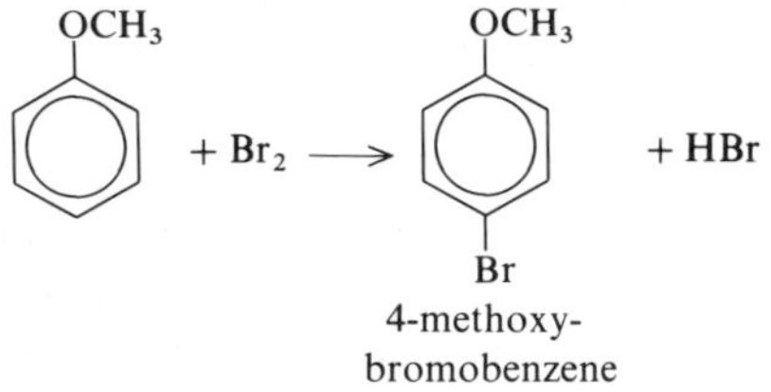

4-methoxy-
bromobenzene

The product will react with more bromine to give

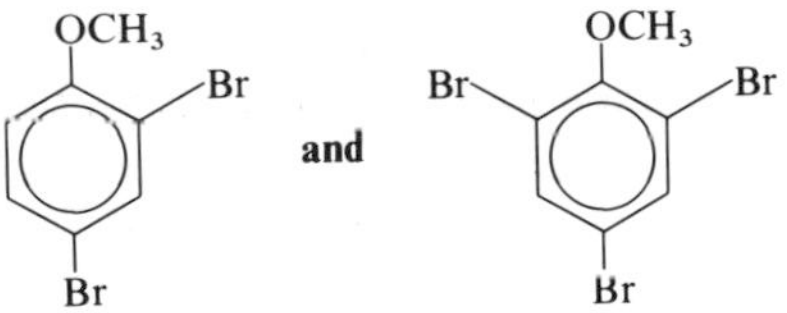

and

with more hydrogen bromide being produced.

These reactions of the benzene ring are known as *substitution reactions*.

EXPERIMENT 9.5b
The nitration of methyl benzoate

As a continuation of your experiments on arenes you may have time to carry out a full scale laboratory preparation as well as the test-tube investigations you have already done. This preparation involves the substitution of a nitro group into a benzene ring.

methyl
benzoate

methyl 3-
nitrobenzoate

Procedure

Wear safety glasses during this experiment.

1 Measure $9 \, cm^3$ of concentrated sulphuric acid (*TAKE CARE*) into a 100-cm^3 conical flask and cool it to below 10°C in an ice bath. Add $4 \, cm^3$ of methyl benzoate while swirling the flask. Prepare a mixture of $3 \, cm^3$ of concentrated nitric acid with $3 \, cm^3$ of concentrated sulphuric acid in a small flask (*TAKE CARE*) and cool the mixture in the ice bath.

2 Use a dropping pipette to add the nitric acid mixture a drop at a time to the methyl benzoate solution. Swirl the conical flask and control the rate of addition so that the temperature stays in the range 5 to 15 °C. The addition should take about 15 minutes.

3 When the addition is complete, remove the flask from the ice bath and allow it to stand at room temperature for 10 minutes. Pour the reaction mixture over 40 g of crushed ice and stir until the product solidifies. Collect the product by suction filtration (wait until all the ice melts). Wash with three portions of water, sucking dry and disconnecting the suction pump before each addition of washing water.

4 Change the Buchner flask for a small clean dry flask and wash the product with two portions of $5 \, cm^3$ of *ice cold* ethanol. Keep the wash liquid for examination by chromatography.

5 To recrystallize the product, transfer it to a 100-cm^3 conical flask and add about $20 \, cm^3$ of ethanol, the minimum volume that will dissolve the solid when hot. Heat a water bath to boiling and turn out the Bunsen burner before putting the conical flask containing the ethanol in the water bath. When the solid has dissolved, it can be recovered by cooling the solution in an ice bath and collecting the crystals which form, by suction filtration. Methyl 3-nitrobenzoate is a pale yellow solid of melting point 78 °C.

6 For chromatography evaporate the wash liquid to $1 \, cm^3$ in an evaporating basin, either by standing it overnight or by heating it on a hot water bath. Use a melting point tube drawn out to a fine tip to put a spot of the solution 2 cm from the bottom of a thin layer of silica on an inert support. Some of the product can be dissolved to make a second separate spot on the plate. Allow the solvent to evaporate and develop with an ethoxyethane–hexane mixture containing 1 volume of ethoxyethane to 9 volumes of hexane (*TAKE CARE: this mixture is highly flammable*).

7 Methyl 2-nitrobenzoate, a minor product, should be visible on the silica sheet as a yellow spot, while methyl 3-nitrobenzoate can be seen under ultra-violet light or by exposing the sheet to iodine vapour. *T A K E   C A R E*: do not look directly at the ultra-violet light.

8 The melting point of the product can now be found.

One form of apparatus used for this purpose is illustrated in figure 9.18. It consists of a hard-glass tube partly filled with dibutyl benzene-1,2-dicarboxylate and holding a thermometer. Put a sample into a small thin-walled capillary tube sealed at one end, and by gentle tapping, or rubbing with the milled edge of a coin, transfer it to the closed end. Fix the tube in the position shown in the figure by means of a rubber band. Slowly heat the tube by means of a very low Bunsen burner flame so as to maintain an even rise of temperature. Watch the crystals in the melting point tube carefully, and the moment they melt, note the temperature. Repeat the process with a fresh melting point tube containing another portion of the compound, in order to obtain a more accurate value for the melting point. The temperature may now be raised rapidly to within 10 °C of the melting point previously obtained, but must then be raised very slowly (about 2 °C rise per minute) until the crystals melt. Note the temperature at which the crystals first melt and also the temperature at which melting is complete. For pure substances these temperatures are close together and the melting point is called 'sharp'.

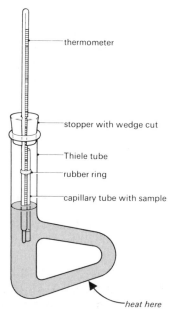

Figure 9.18
The Thiele melting point apparatus.

If the compound under examination is then recrystallized and dried, and the melting point again determined, it may be found to be a little higher than before. This is because the melting point of a pure compound is always lowered by the presence of impurities. The compound can be made completely pure by repeated recrystallization until the melting point is constant.

An interpretation of the substitution reaction of the benzene ring

We can now consider how the bromine substitution reaction of the benzene ring takes place.

We have already said that this type of reaction, in which a hydrogen atom of a benzene ring is replaced by another atom, is known as a substitution reaction. You have seen that it takes place quite easily with methoxybenzene but not with methylbenzene. Methylbenzene will undergo such a reaction but a catalyst (iron is suitable) is needed.

$$\text{C}_6\text{H}_5\text{CH}_3 + \text{Br}_2 \xrightarrow[\text{catalyst}]{\text{Fe}} \text{CH}_3\text{C}_6\text{H}_4\text{Br} + \text{HBr}$$

The major product is the monobromo- compound, although the yields of dibromo- and tribromo-methylbenzene can be increased by heating.

What evidence is there of the nature of the attacking group? It has been found that the reaction of iodine monochloride, I—Cl, with methoxybenzene produces only iodine substitution products.

$$\text{C}_6\text{H}_5\text{OCH}_3 + \text{I}-\text{Cl} \longrightarrow \text{CH}_3\text{O}\text{C}_6\text{H}_4\text{I} + \text{HCl}$$

What is the attacking atom in this reaction? What polarization would you expect in I—Cl? So what is the charge on the attacking atom in the reaction?

Now consider the leaving group. What atom is lost from the benzene ring in the reaction? Will this atom more easily carry a positive or a negative charge when it leaves the benzene ring? Is this consistent with the charge which, you have suggested, the attacking atom will bring to the benzene ring?

You should now have a hypothesis about the charge on the attacking agent in a benzene ring substitution and also about the nature and charge of the leaving group. We can see if this hypothesis is consistent with the relative ease of attack on methylbenzene and methoxybenzene.

What polarization of the benzene ring is required to facilitate attack by the iodine atom of iodine monochloride? The polarization of the benzene ring caused by substituents will be indicated by the dipole moment of the molecules.

Molecule	Direction of dipole	Dipole moment/D
benzene ring—OCH_3	$\leftarrow +$	1.38
benzene ring—CH_3	$\leftarrow +$	0.36
benzene ring		0.0

Do the dipole moments change in parallel with the reactivity of the benzene ring? If you examine an electron cloud model of methoxybenzene you will see that the p-electrons on the oxygen are available to interact with the delocalized π-electrons in the benzene ring. This is considered to be the source of the greater reactivity of methoxybenzene. Check that this theory is consistent with your hypothesis about the nature of the attacking group.

Finally, let us examine the function of the iron catalyst. Iron reacts with bromine to form iron(III) bromide:

$$2Fe + 3Br_2 \longrightarrow 2FeBr_3$$

This in turn induces polarization in other bromine molecules:

$$FeBr_3 + Br_2 \longrightarrow Br^{\delta+}\!-\!Br^{\delta-} \cdot FeBr_3$$

Reaction of this last compound with methylbenzene regenerates the iron(III) bromide, and the catalyst is therefore iron(III) bromide and not iron.

Thus, the function of the catalyst is to provide a bromine atom carrying the correct charge for attack on the benzene ring.

Reactions of the benzene ring

We can now see that the special reaction of the benzene ring can be described as an *electrophilic substitution,* with electron-deficient reagents attacking the benzene ring. The benzene ring can donate a pair of electrons to the attacking group. This theory can be enlarged to interpret the positions on the benzene ring that are attacked, but we shall not be following the theory as far as that.

1 *Halogenation* Bromine, usually in the presence of a catalyst, such as iron(III) bromide to make the bromine molecules more electrophilic, substitutes a bromine atom for a hydrogen atom.

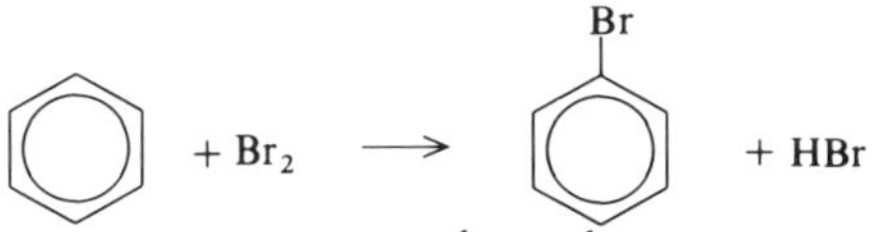

bromobenzene

2 *Sulphonation* Fuming sulphuric acid is used for sulphonation, giving products which are often water-soluble. Refluxing for several hours is often necessary. The electrophile is considered to be sulphur trioxide, SO_3.

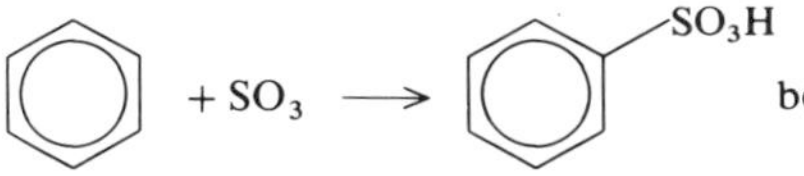

benzenesulphonic acid

In this reaction, benzene gives benzenesulphonic acid. This compound, like sulphuric acid, ionizes in water.

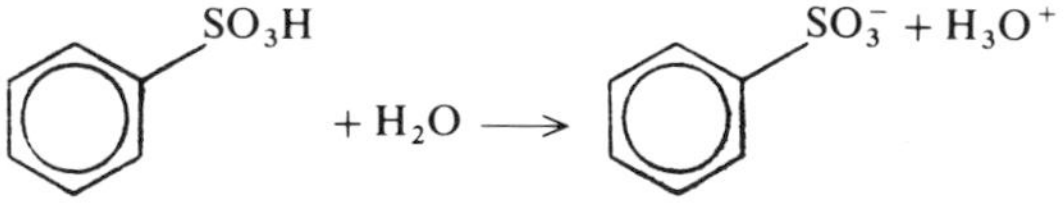

Sulphonation is used in the manufacture of a wide range of substances including sulphonamide drugs, detergents, and dyestuffs.

3 *Alkylation by the Friedel–Crafts reaction* Chloroalkanes, RCl, in the presence of aluminium chloride, will form a complex, $R^+AlCl_4^-$, in which R^+ acts as an electrophile (R represents any alkyl group).

$$CH_3CH_2Cl + AlCl_3 \longrightarrow CH_3CH_2^+ AlCl_4^-$$

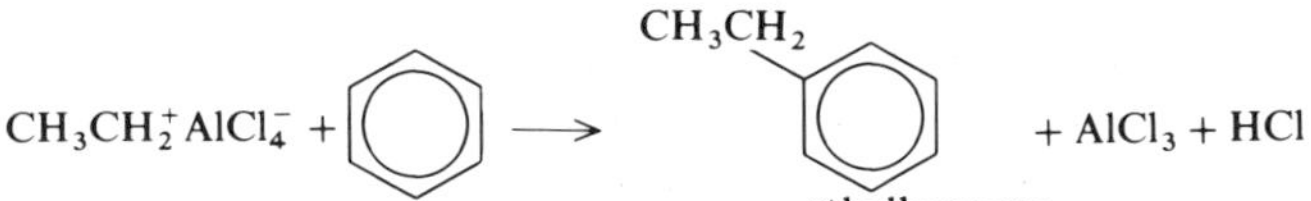

ethylbenzene

The aluminium chloride is a catalyst so only small quantities are needed to carry out the reaction.

This reaction has important industrial applications.

4 *Nitration* Nitric acid in the presence of concentrated sulphuric acid produces the electrophile NO_2^+. The reaction of benzene with the electrophile NO_2^+ substitutes a nitro group NO_2 for a hydrogen atom.

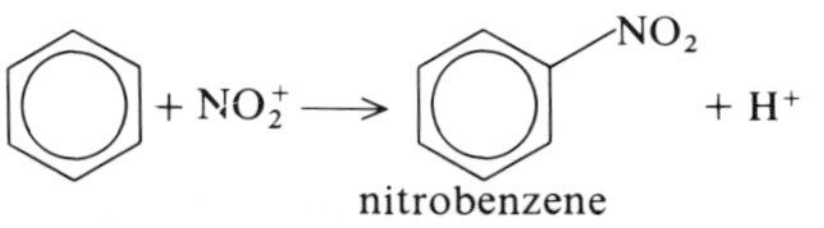

nitrobenzene

Reactions of this type are known as *nitrations*. They are used in the manufacture of explosives (such as TNT, trinitrotoluene) and dyestuffs.

5 *Addition reactions of benzene* In severe conditions benzene will undergo some addition reactions. Thus hydrogen in the presence of a nickel catalyst will react to form cyclohexane. A temperature of 200 °C is necessary, and a pressure of 30 atmospheres is used to keep the reaction in the liquid phase. This reaction is the main source of the high purity cyclohexane needed for the manufacture of nylon (see Topic 17).

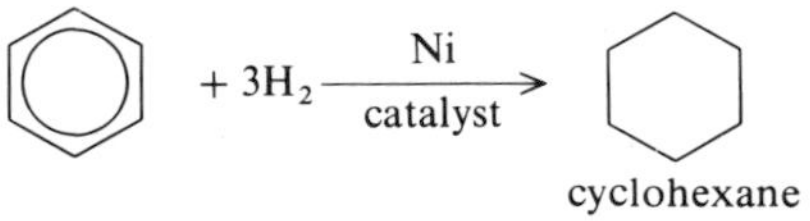

cyclohexane

Chlorine will also add to benzene when irradiated with ultra-violet light, and this is another example of a free radical reaction. The details of this reaction are given in the Background reading on insecticides that follows.

BACKGROUND READING
Insecticides based on benzene

The benzene ring forms part of the molecular structure of many useful compounds including a number of insecticides. Two examples of wellknown insecticides with structures containing benzene rings are BHC and DDT.

BHC (benzene hexachloride)

This compound is made by passing chlorine through liquid benzene irradiated by ultra-violet light.

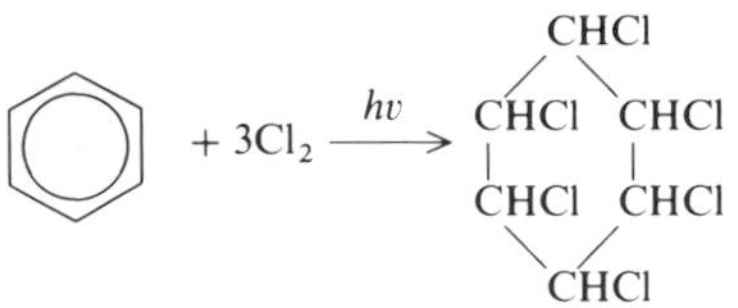

1,2,3,4,5,6-
hexachlorocyclohexane

BHC is particularly valuable in the fight against the locust. The devastation caused by locust swarms can be judged from figure 9.19.

DDT (dichlorodiphenyltrichloroethane)

This is made by a reaction between chlorobenzene and trichloroethanal (chloral). It has been used extensively against mosquitoes in an effort to eliminate malaria from various regions – in particular, from Sicily and Southern Italy.

The equation for the manufacture of DDT is as follows:

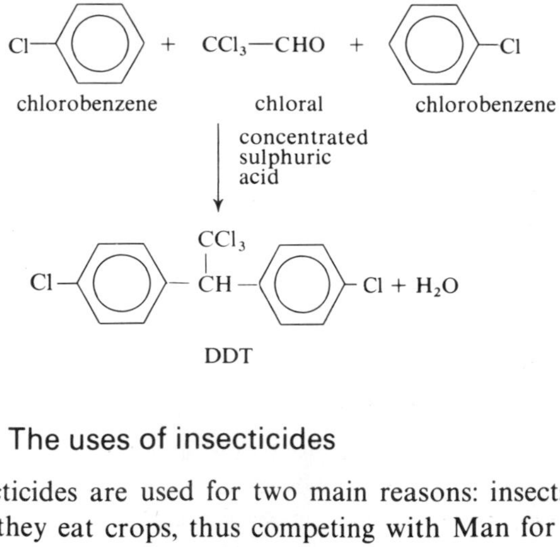

The uses of insecticides

Insecticides are used for two main reasons: insects transmit serious diseases, and they eat crops, thus competing with Man for the World's available food resources. A number of insecticides are not selective in their action, since they kill both harmful and beneficial insects, as well as birds and small animals. Their use is controlled in many parts of the World. This is doubly necessary because misapplication can have serious secondary effects. DDT is a stable chemical and persistent in the environment, so that when rain falls on treated land the DDT may be washed into streams and lakes. Aquatic animals are readily able to absorb chemicals, including DDT, from the water. Since DDT is much more soluble in oils and fats than it is in water, it tends to be retained in animal bodies for a long time, and long-lived animals such as fish can build

Figure 9.19
The devastation caused by locusts. The first photograph shows a typical productive orange grove near Tripoli, Libya. The second, a similar orange grove after a locust attack in the Souss Valley, Libya.
Photographs, Shell.

up damaging amounts of it in their bodies. DDT can enter the bodies of fish not only when they absorb it directly from the water but also when they feed on other animals which have themselves absorbed it.

Against the risks, we must set the benefits arising from the use of insecticides. Pain and suffering are lessened and the expectation of life is increased because diseases caused by insect carriers are reduced. Chemical means of protection result in vast increases in crop production. Nevertheless it is difficult to weigh the short term advantages for today's world against the possible long term disadvantages that future generations may have to struggle to put right. The particular problem with BHC and DDT is that they are very resistant to degradation, in the environment, to harmless end-products.

DDT is no longer much used for agricultural pest control, having been replaced by more readily degradable chemicals, such as organic phosphorus compounds, carbamates, and synthetic pyrethroids. Some of these are much more selective than DDT and less damaging to animals which are not pests. Many of these newer insecticides, such as Permethrin, also contain the benzene structure.

Permethrin

DDT is, however, a cheap and effective insecticide and is still retained for use in malaria control, where a persistent material is required to go on killing malaria-carrying mosquitoes for several months after one application.

In tropical climates especially, insect damage to stored crops can be considerable. For example in Kano, Nigeria, it was found in one experiment that when two water traps, each about $250 \, cm^2$, were placed in and near the doorway of a groundnut store, a catch of eight hundred groundnut beetles, *Tribolium castaneum*, was made in twenty-four hours.

Another study carried out in Zimbabwe evaluated damage to beans. These are an important crop in that country, but in storage the beans are often much damaged by beetles belonging to the family Bruchidae. G. F. Cockbill, an entomologist, was asked therefore to find out if the application of a BHC dust would be an economical method of reducing damage.

Two stacks of beans, each of a hundred and five bags in layers of five to three bags, were prepared. One stack was used as a control and left untreated. The other stack was treated with a dust containing 0.65% gamma BHC, each layer of bags receiving dust at the rate of one gram per square metre.

Before building the stacks twelve bags were selected at random, the beans were weighed, and a sample was removed to assess the content of moisture and the damage done by insects. Six of the sampled bags were used in the building of each stack, three bags occupied outside positions in different layers, and three had central positions.

The stacks were left undisturbed for five months in a covered store, so in assessing the losses due to insect attack, a net loss in mass was determined which allowed for variation in moisture content.

	% loss in mass in control stack	% loss in mass in treated stack
outside bags	13.0	1.5
central bags	13.5	5.5
mean	13.3	3.5

Table 9.1
Loss in mass in beans affected by Brucid beetles.

The results indicated that treatment saved approximately 10% of the total quantity of beans from damage by insects. In the control stack, position in the stack gave no protection, although in the treated stack, the central bag from the middle layer (one of the few bags not partially exposed) was one of the least attacked; it suffered only a 0.7% net loss in mass.

The number of beans damaged was also determined. This affects the eating quality because in heavy attack little is left of a bean but its tough seed coat.

	% increase in damaged beans In control stack	% increase in damaged beans in treated stack
outside bags	41.7	7.0
central bags	17.3	16.7
mean	29.5	11.8

Table 9.2
Increase in damage in beans attacked by Brucid beetles.

The mean results indicate that during the five months of the experiment the number of damaged beans in the control stack increased by about 30% but treatment kept the increase in damage to about 12%.

The effectiveness of the BHC treatment in killing Brucid beetles was found by recovering and counting beetles from the sample sacks. A mean of fifteen dead beetles was found in sacks from the control stack and twice as many in the treated sacks.

As a final test, samples of beans were prepared and cooked. No unpleasant odour or flavour was detected in the treated beans. These were preferred 'because there were fewer skins'.

The cost of treatment at the time was 3p per hundred bags and it produced a saving of about 10% in a crop valued at £300 per hundred bags. Thus, the treatment was concluded to be effective, economical, and practicable.

9.6
SURVEY OF REACTIONS AND REAGENTS IN TOPIC 9

This Topic has introduced you to a number of important ideas that you need to understand before you proceed to Topic 11, the next one on the organic chemistry of carbon. You will also need to learn the particular reagents, and conditions, for each reaction. When this basic information has been learned you should make sure you understand the interpretations of the reactions and that you are familiar with at least some of the Background reading included in the chapter.

These lists should be treated as statements to be understood, not definitions to be memorized. You should find an example of your own choice for each item, to help you to understand the ideas involved.

Types of reaction

A *substitution reaction* is a reaction in which one group replaces another in a molecule.

$$CH_3CH_2Br + NaOH \longrightarrow CH_3CH_2OH + NaBr$$

An *addition reaction* is one in which one or more groups are added onto a molecule, to give a single product.

$$CH_2{=}CH_2 + Br_2 \longrightarrow CH_2Br{-}CH_2Br$$

An *elimination reaction* is one in which one or more groups are removed from a molecule.

$$CH_3CH_2Br \longrightarrow CH_2{=}CH_2 + HBr$$

Notice that this is the reverse of an addition reaction.

A polymerization reaction is one in which molecules with a small molecular mass join up to become molecules with a large molecular mass.

$$nCH_2{=}CH_2 \longrightarrow +CH_2{-}CH_2{+}_n$$

A chain reaction is one in which molecules of product are produced at each cycle of a process that usually repeats itself a large number of times.

$$CH_4 \xrightarrow{\;\cdot Cl\;} \cdot CH_3 + HCl \xrightarrow{\;Cl_2\;} CH_3Cl + \cdot Cl$$

This process includes the stages of initiation, propagation, and termination.

Bond breaking

Homolytic bond breaking involves the breaking of a bond so that the electrons are equally shared between the two atoms, or free radicals.

$$Cl{-}Cl \longrightarrow Cl\cdot + Cl\cdot$$

Heterolytic bond breaking involves the breaking of a bond so that the electrons are unequally distributed between the two atoms, forming ions.

$$H{-}Cl \longrightarrow H^+ + :Cl^-$$

Types of reagent

Nucleophiles are attacking groups with a pair of electrons available for forming a new covalent bond. They are often negatively charged.

$$:OH^-, \quad :NH_3, \quad :CN^-$$

Electrophiles are attacking groups with a vacancy for a pair of electrons; a new covalent bond results when the vacancy is filled. Electrophiles are often positively charged.

$$H^+, \quad NO_2^+, \quad SO_3, \quad R^+(AlCl_4^-)$$

Free radicals are uncharged attacking groups with an odd number of electrons, so they possess only one of the electron pair needed for the formation of a new covalent bond.

$$\cdot Cl, \quad \cdot CH_-$$

Reactions of alkanes

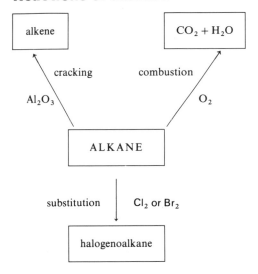

Reactions of halogenoalkanes

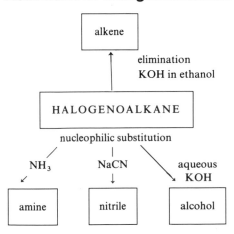

Reactions of alkenes

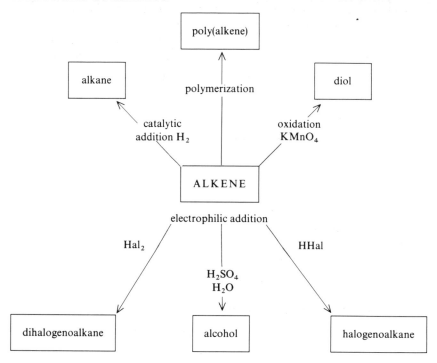

Reactions of benzene

Using the three examples on page 321 and above, draw up your own chart for benzene. You can include more (or less) information, depending on how helpful you find charts as a method of learning.

SUMMARY

At the end of this Topic you should:

1 be aware of the scope of carbon chemistry, the types of molecular structures to be found, and the ways in which these structures can be represented, both as three-dimensional models and in two dimensions on paper;

2 understand, and be able to use, the systematic methods of nomenclature used for the carbon compounds that have been described in the Topic;

3 be familiar with some organic practical procedures;

4 understand the types of reaction, and reagents, summarized in the preceding section (page 319);

5 know the chemical reactions of the various classes of compounds that are given in the parts of the Topic headed 'Reactions of the ...' and summarized in the preceding section (page 319);

6 be aware of the uses of organic compounds as anaesthetics and as insecticides, and be aware of the use of alkanes in petrol, including the meaning of octane number, and its relationship with structure.

PROBLEMS

*Indicates that the *Book of data* is needed.

Alkanes

1 Draw the structural formula of each of the following:

a 2-methylbutane
b 2,2-dimethylpropane
c 2,3,3-trimethylpentane
d 3-ethyl-4,4-dimethylheptane

2 Name the following compounds.

a $CH_3CH_2CH_2CH_2CH_3$

b $CH_3CHCH_2CH_2CH_3$
$\qquad\;\;|$
$\qquad CH_3$

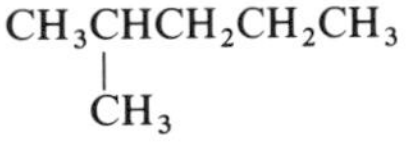

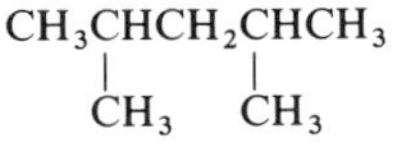

c $CH_3CHCH_2CHCH_3$
$\qquad\;\;|\qquad\quad|$
$\qquad CH_3\quad CH_3$

d $CH_3CH{-}CH{-}CHCH_3$
$\qquad\quad|\quad\;\;|\quad\;\;|$
$\qquad CH_3\; CH_3\; CH_3$

$\qquad\qquad CH_3$
$\qquad\qquad |$
e $CH_3CH_2CCH_2CH_3$
$\qquad\qquad |$
$\qquad\qquad CH_3$

f $\quad CH_3$
$\qquad |$
$CH_3CCH_2CHCH_3$
$\qquad |\qquad\quad|$
$\qquad CH_3\; C_2H_5$

g $\qquad CH_2$
$\qquad\nearrow\quad\searrow$
$\quad CH_2\qquad CH_2$
$\quad |\qquad\qquad|$
$\quad CH_2{-}CH_2$

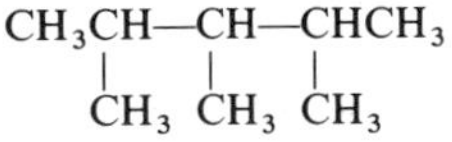

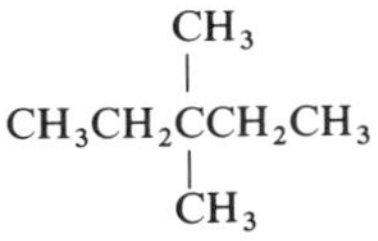

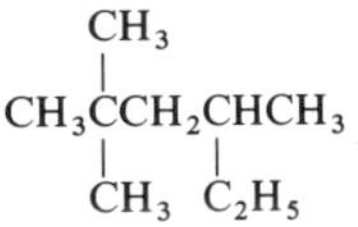

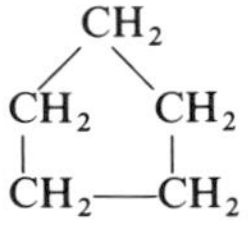

3 Write down the structures and names for all the possible hexanes C_6H_{14}.

*4a Use the *Book of data* to find the enthalpy change of formation of methane and the enthalpy changes of atomization of carbon and hydrogen.

Determine the enthalpy change of formation of methane from free gaseous atoms. From this, calculate the average bond energy of the carbon–hydrogen bond.

b Use the result you obtained in **a**, together with the appropriate data for ethane, to calculate the bond energy of the carbon–carbon bond in ethane.

Halogenoalkanes

5 Write the structural formulae of:

a 2-bromo-2-methylpropane
b 4,4-dichloro-3-ethylhexane
c 1,2,3,4,5,6-hexachlorocyclohexane (an important insecticide).

6 Name the following compounds:

a $C_2H_5Cl(CH_3)CH_2CHClCH_3$
b $CH_3CCl_2C(CH_3)_2CH_2CH_3$
c $CHBrClCF_3$ (an important anaesthetic).

7 Write down the structures and name all the compounds with the molecular formula C_4H_9Br.

8 Explain how, starting with the appropriate mono- or dihalogenated alkane, you might synthesize the following:

a $CH_3CHOHCH_3$ (propan-2-ol)
b $H_2N(CH_2)_6NH_2$ (1,6-diaminohexane)
c CH_2CO_2H (butanedioic acid)
$\quad|$
CH_2CO_2H
(*Hint:* Nitrile groups, —CN, can be converted to the corresponding acid groups, —CO_2H, by boiling with mineral acids.)
d $CH_2{=}CHCH_2CH_3$ (but-1-ene).

9 The following scheme shows some reactions of 1-bromopropane.

a State the conditions necessary to convert 1-bromopropane into propene (I).

b State the conditions necessary to convert 1-bromopropane into propan-1-ol (II).

c Give the structural formulae of substances (III), (IV), and (V). Explain the importance of this sequence of reactions.

10a Draw the electronic structure (outer shell only) of the water molecule, H_2O.

b Draw the electronic structures of the *atom* and *free radical* that would be obtained if one of the O—H bonds in a water molecule underwent *homolytic fission*.

c Draw the electronic structure of the *ions* that would be obtained if one of the O—H bonds in a water molecule underwent *heterolytic fission*.

d Explain why the ion OH⁻ is said to be nucleophilic.

e Can you suggest a similar name which might be appropriate to describe the ion H^+?

11 Choose from the following list those reagents which you think might possess nucleophilic properties.

H^+, CN^-, $CH_3\cdot$, OH^-, Cu^{2+}, H_2O, NH_3, $Br\cdot$, Br^-, CH_3NH_2, Na^+

12 Halogenoalkanes can undergo both nucleophilic substitution reactions and elimination reactions with the same reagent; this is because nucleophiles can also act as bases. Elimination, which depends upon the removal of a proton from the halogenoalkane, is generally more difficult because it depends upon breaking the very strong C—H bond: furthermore, it is assisted by reducing the polarity of the solvent.

Which one of the following ways of reacting the halogenoalkane would be most likely to favour elimination?

A with a strong base, at a low temperature in aqueous solution

B with a weak base, at a high temperature in aqueous solution

C with a strong base, at a low temperature in solution in ethanol

D with a strong base, at a high temperature in solution in ethanol

E with a weak base, at a high temperature in solution in ethanol.

Alkenes

13 Write structural formulae and give the names for as many compounds as possible with the molecular formula C_4H_8.

14 Which of the following substances can exist as geometric isomers (*cis*- and *trans*- forms)?

$(CH_3)_2C{=}CH_2$ $CH_3CH{=}CHCH_3$ $ClCH{=}CCl_2$

     A   B   C

$CH_3CH{=}CHCH_2CH_3$

     D

15 Which of the following molecules would you expect to have a dipole moment?

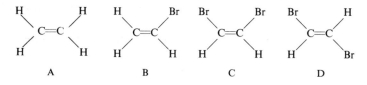

    A   B   C   D

16 Which of the following groups of reagents all form stable addition products with ethene, with or without the use of catalysts?

A $Br_2(l)$, $NaOH(aq)$, $H_2(g)$, $H_2SO_4(aq)$
B $H_2SO_4(aq)$, $NH_3(aq)$, $HBr(g)$, $C_2H_4(g)$
C $HBr(g)$, $H_2(g)$, $Br_2(l)$, $C_2H_4(g)$
D $CuSO_4(aq)$, $H_2(g)$, $H_2SO_4(aq)$, $HBr(g)$
E $C_2H_4(g)$, $Br_2(l)$, $CuSO_4(aq)$, $H_2O(l)$.

17 Give equations, reagents, and reaction conditions to show how you would perform the following syntheses:

a CH_3CH_2Br to CH_2BrCH_2Br (2 steps)
b $CH_2{=}CH_2$ to CH_3CH_2CN (2 steps)
c $CH_2{=}CH_2$ to CH_2OHCH_2OH (2 steps)

18 Classify the following species as free radicals, electrophiles, or nucleophiles, giving reasons for your choice by indicating the number of electrons in the outer shell of the significant atom.

Br^+, OH^-, $CH_3\cdot$, H_2O, CH_3^+, $Cl\cdot$, I^-, H^+, CH_3NH_2, BF_3

Why are there no metal cations such as Na^+ or Cu^{2+} in the list, although anions of non-metals are represented?

19 Bromine reacts readily with alkenes in a glass flask, but if the flask is internally coated with paraffin and all the reagents are perfectly dry, the reaction will not take place.

On the basis of this evidence alone, which of the following deductions can you make?

A the reaction will only take place in the presence of moisture
B an alkene π bond must be polarized by an electrophile before addition can take place
C the presence of some kind of polar material is necessary to initiate the reaction
D paraffin wax inhibits addition reactions.

20 Which of the substances listed below would be formed in the reaction of ethene with aqueous bromine in a saturated solution of sodium chloride? (More than one of these substances may be formed.)

CH_2BrCH_2Br, CH_2BrCH_2Cl, CH_2ClCH_2Cl, CH_2BrCH_2OH,
    A   B   C   D

CH_2OHCH_2OH
    E

What are the implications of your choice for the mechanism of the reaction?

Arenes

21 Copy out and complete the following table, using structural formulae instead of names for the stated substances.

Starting material	Benzene	Cyclohexene	Cyclohexane
Product	Bromobenzene	1,2-dibromocyclohexane	1-bromocyclohexane
Reagent used	Bromine	Bromine	Bromine
Essential conditions for reaction to take place			
Type of reaction (addition, substitution, elimination)			
Type of reagent (electrophile, nucleophile, free radical)			

22 What reactions (if any) might be expected to take place between benzene and the following reagents, assuming that appropriate catalysts may be used? Give reasons for your choice and state which catalysts would be used.

a NH_3 **b** $BrCl$ **c** KOH **d** CH_3CH_2I **e** DCl (D = deuterium ^2H)

*23 The diagram below shows some reactions of methylbenzene (see I):

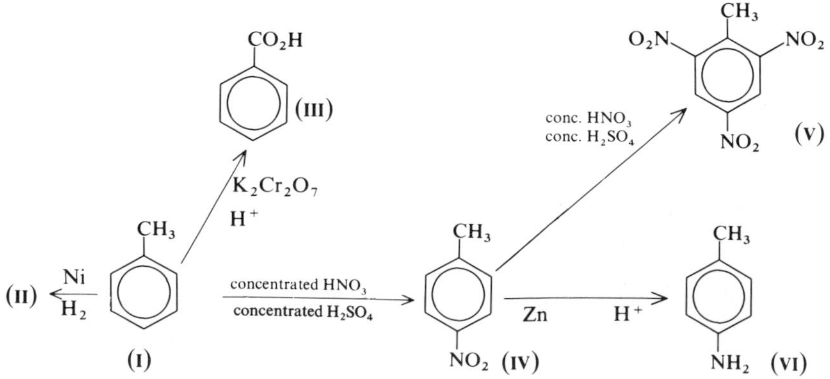

a Give the name and structure of product (II).

b Classify the following reactions as oxidation, reduction, substitution, addition, or elimination.

(I) $\longrightarrow$ (II)   (I) $\longrightarrow$ (III)   (I) $\longrightarrow$ (IV)   (IV) $\longrightarrow$ (V)   (IV) $\longrightarrow$ (VI)

  A   B   C   D   E

c What characteristics of product (v) lead to its use as an important explosive (TNT)?

d Outline the procedures of an experiment which would enable you to distinguish methyl-4-nitrobenzene, product (IV), from the other two isomers which might have been formed, using the technique of thin layer chromatography.

e Use the *Book of data* to suggest how substance (III), benzoic acid, could be identified, using infra-red spectroscopy.

24 Chlorine reacts with boiling benzene in sunlight to give a mixture of several isomers, all with the molecular formula $C_6H_6Cl_6$, which are useful as insecticides.

a What type of reaction has taken place and what is the attacking species?

b Why is this type of reaction unusual for benzene?

c Suggest how the isomers differ from one another in structural terms.

*25 Starting from the standard enthalpy change of formation of benzene, calculate its heat of atomization – that is, the enthalpy change for the process:

$$C_6H_6(g) \rightarrow 6C(g) + 6H(g)$$

Also, use the average bond energies in the *Book of data* to calculate this same value for the gaseous molecule

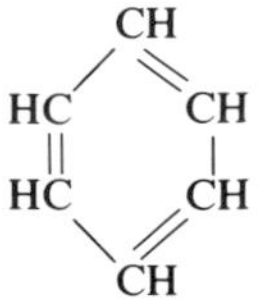

```
        CH
      /     ⟍
   HC        CH
   ‖          |
   HC        CH
      ⟍     ⟋
        CH
```

(The enthalpy change of evaporation of benzene is:
$\Delta H_e = +33.9 \, \text{kJ mol}^{-1}$.)

Comment on the values you obtain.

*26a The enthalpy change of hydrogenation of cyclohexene is $-120\ \text{kJ mol}^{-1}$.
Assuming the structural formula of naphthalene to be

what would you expect its enthalpy change of hydrogenation to be?

b Making use of the average bond energies in the *Book of data*

$E(C—C)$ general, $E(C{=}C)$ general, $E(C—H)$ general, $E(H—H)$

calculate a second value for the enthalpy change of hydrogenation of
naphthalene, assuming it to have the structural formula given above.

c What difference is there between your answers to **a** and **b** and how may
it be explained?

d How would you expect the experimental enthalpy change of
hydrogenation to compare with the values you have determined and how
may any differences be explained?

Liquid–Vapour Equilibria

In this Topic we shall investigate the vapour pressures that exist above single liquids, and see how they are modified

1 by changes in temperature, and

2 by the presence of solutes, both involatile and volatile.

We shall see how our results enable us to separate liquids by distillation, and to determine the relative molecular masses of solutes. We shall also meet a molecular picture of the events described, using the idea of entropy developed in earlier Topics.

The study of vapour pressures will lead us to a study of the forces of attraction that exist between molecules. These intermolecular forces affect physical properties to a considerable extent, and can be vital in promoting essential biochemical reactions. Amongst many other things, their study leads us to understand why chlorine is a gas and iodine is a solid, and why water is a liquid but hydrogen sulphide is a gas, and it contributes to our understanding of the structure of proteins.

10.1
ENTHALPY AND ENTROPY CHANGES ON VAPORIZATION

We shall begin our study of liquid–vapour equilibria by finding out the enthalpy change that takes place when water is boiled. We shall then use this value to calculate the corresponding entropy change.

EXPERIMENT 10.1
To determine the enthalpy change, ΔH_b, and entropy change, ΔS_b, for the vaporization of water

Procedure

Set up the apparatus as shown in figure 10.1. Have a second, weighed, 100-cm³ beaker ready. Switch on the electrical supply to the meter and wait until water is distilling steadily. Switch off the electricity, read the joulemeter, and place the weighed beaker to collect the distillate. Switch on the electricity and distil over approximately 10 cm³ of water. Switch off, reweigh the beaker, and read the joulemeter.

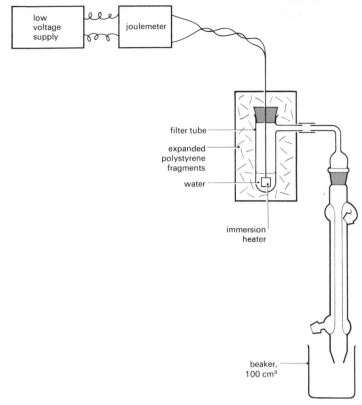

Figure 10.1

Calculations

1 Calculate the mass of water distilled. What is this amount in moles?

2 Calculate the enthalpy change of vaporization (sometimes called the molar latent heat of vaporization) from the equation

$$\Delta H_b / \text{kJ mol}^{-1} = \frac{\text{electrical energy supplied/kJ}}{\text{amount of water distilled/mol}}$$

3 Calculate the entropy change of vaporization from the equation

$$\Delta S_b / \text{kJ mol}^{-1}\,\text{K}^{-1} = \frac{\Delta H_b / \text{kJ mol}^{-1}}{\text{boiling point/K}}$$

(The boiling point of water, T_b, can be taken as 373 K.)

4 As the numerical value of ΔS_b is low in these units, convert your answer to J mol^{-1} K^{-1}, the usual units of entropy. 109 J mol^{-1} K^{-1} is the accepted value.

Patterns in entropies of vaporization

Using modified versions of the apparatus shown in figure 10.1, it is possible to find the enthalpy changes of vaporization of a large number of substances. Some typical results are given below

Substance	Formula	ΔH_b/kJ mol^{-1}	T_b/K
Carbon disulphide	CS_2	27.2	319
Trichloromethane	$CHCl_3$	29.3	335
Tetrachloromethane	CCl_4	30.4	350
Benzene	C_6H_6	30.9	353
Methanol	CH_3OH	35.2	338
Ethanol	C_2H_5OH	38.5	352
Hexane	C_6H_{14}	28.8	342
Octane	C_8H_{18}	34.9	399
Methylbenzene (toluene)	C_7H_8	33.4	384
Cyclohexane	C_6H_{12}	30.1	354
Methylcyclohexane	C_7H_{14}	31.7	374
Ethanoic acid	CH_3CO_2H	24.3	391
Water	H_2O	40.6	373
Sulphuric acid	H_2SO_4	50.2	617
Mercury	Hg	59.1	630

Table 10.1

Plot a graph of ΔH_b (y axis) against T_b (x axis), putting the formula of the substance beside the appropriate point. Most of the points can be joined by a straight line passing through the origin. Draw this line on the graph. Find the gradient of this line; this is $\dfrac{\Delta H_b}{T_b}$, and so is the entropy change of vaporization for those liquids whose points lie on the line. Make a list in your notebook of those liquids which have entropies of vaporization much greater or smaller than this.

The entropy change of vaporization: a simple case of equilibrium

Since ideal gases all occupy the same molar volume, and since molar volumes of liquids are all much smaller, we will not be far wrong if we say that many substances increase volume by much the same amount when they vaporize. This means that the entropy increase, which is mainly due to all the extra space the molecules have to occupy, and to move about in, may well be much the same for many liquid–gas changes.

As we know, liquids left in the open generally evaporate. This is because the entropy increases when the molecules occupy more space. Furthermore, any breeze there may be keeps the amount of vapour near the liquid small, so that locally the partial pressure is lowered. This helps by increasing the effective volume still more.

But in a closed vessel there is an equilibrium. At equilibrium, if an amount of liquid turns into vapour, causing an increase in entropy, an exactly equal amount of vapour will turn into liquid, reducing the entropy to its former value. Both processes happen to the same extent either way, so neither can increase or decrease the entropy. More formally, at equilibrium

$$\Delta S_{\text{total}} = 0$$

In Topic 6 we found how to calculate the entropy change when molecules take energy from the random thermal motion of others. We use the relationship

$$\Delta S_{\text{surroundings}} = -\Delta H/T$$

The boiling point of a liquid is the temperature of the surroundings below which the vapour condenses and above which the liquid evaporates. At the boiling point we shall have

$$\Delta S_{\text{total}} = 0$$

Suppose the entropy increase due to the increase in volume (and anything else, such as an increase in the number of ways that the molecules can vibrate) is ΔS_{evap}. Then we expect

$$\Delta S_{\text{evap}} - \Delta H_{\text{b}}/T_{\text{b}} = 0$$

Now, let us suppose that the increase in the volume in which the molecules can move is the major factor in ΔS_{evap}. As this increase in volume is much the same for many substances then it is likely that ΔS_{evap} will not vary very much from one liquid to another. If this is right, then $\Delta H_{\text{b}}/T_{\text{b}}$ will have approximately the same value for all liquids. This statement, that $\Delta H_{\text{b}}/T_{\text{b}}$ is approximately the same for all liquids, is known as *Trouton's Rule*.

The graph that you obtained by using the data in table 10.1 shows that the rule holds quite well. When it does not hold there is usually a good reason. The molecules may, for example, be large and flexible, in which case ΔS_{evap} is bigger than expected (as we thought only of the space in which they move) so that $\Delta H_{\text{b}}/T_{\text{b}}$ is bigger than we expect by the rule. Alternatively, the liquid may have some organization or pattern amongst its molecules, in which case again the ratio is bigger.

The derivation of Trouton's Rule illustrates an important idea: to decide about a chemical change, we must consider *all* the ways in which molecules affected by the change can be arranged or have energy arranged amongst them. This means including the surroundings too.

10.2
CHANGE OF VAPOUR PRESSURE WITH TEMPERATURE

Figure 10.2 shows a closed flask containing some water. The air has been pumped out. Water molecules leave the liquid to form water vapour in the space above the liquid. As the number of water molecules in this space increases, some of them lose energy by collision with each other and with the sides of the flask and fall back into the liquid. Very soon an equilibrium is set up between the water molecules in the liquid and those in the vapour:

$$H_2O(l) \rightleftharpoons H_2O(g)$$

The molecules in the vapour exert a pressure by bombarding the sides of the flask. This pressure is known as the vapour pressure of the liquid.

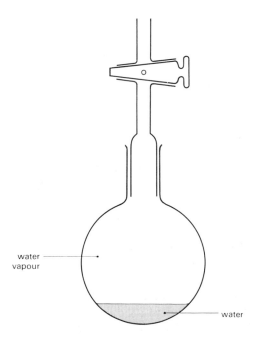

water vapour

water

Figure 10.2

EXPERIMENT 10.2 (demonstration)
To investigate the variation, with temperature, in the vapour pressure of water

The vapour pressure of water varies with the temperature, and the way in which it does so can be investigated, by using the apparatus shown in figure 10.3.

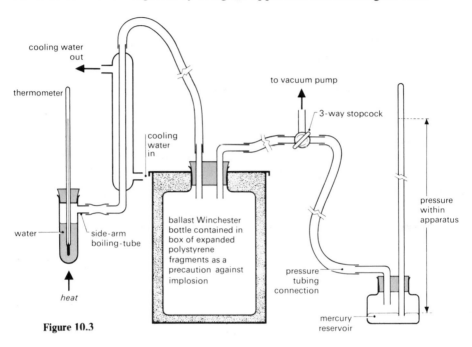

Figure 10.3

In this experiment the pressure above the water is measured. The water is then heated, and the temperature at which it boils is recorded. The pressure is then reduced and measured, and the new boiling point is found; this process is repeated several times. The water boils when its vapour pressure becomes equal to the atmospheric pressure above it. We therefore know the vapour pressure of the water (because it is equal to the measured atmospheric pressure) and the temperature (because this is the boiling point at that pressure). By recording the variation of boiling point with pressure, we are able to see the variation of vapour pressure with temperature.

You may be shown apparatus of this type and be able to take your own readings; if not, you should use the values given in the table, which were obtained in such an experiment.

It is usual in this type of apparatus to measure pressure in the practical units of millimetres of mercury, although it is recognized that these units are not dimensionally correct. To convert such values to the SI pressure unit, the kilopascal, multiply them by 0.133.

Temperature/°C	Temperature/K	Vapour pressure/mmHg	Vapour pressure/kPa
101.0	374.0	770	102.7
97.5	370.5	675	89.9
96.0	369.0	625	83.3
93.0	366.0	573	76.4
91.0	364.0	528	70.4
88.0	361.0	472	62.9
82.0	355.0	376	50.1
79.0	352.0	326	43.5
74.0	347.0	271	36.1
69.0	342.0	192	25.6
61.0	334.0	157	20.9

Plot a graph of vapour pressure/kPa against temperature/K. Use the graph to determine the boiling point of water under standard pressure, *i.e.* 101 kPa. How does the vapour pressure of the water vary with temperature?

Entropy and the change of vapour pressure with temperature

In Topic 3 we saw that the entropy of a gas was larger if its pressure were reduced (if it had more space to move in). This means that a vaporizing liquid can 'afford' the larger entropy drop, $-\Delta H/T$, of the surroundings at a *lower* temperature, if the pressure of the vapour is lower.

This is why vapour pressure falls with temperature.

The actual relationship between vapour pressure and temperature is

$$Lk \ln p = \text{constant} - \Delta H/T$$

You may be interested to see how this relationship can be obtained, though you will not be expected to be able to derive it for yourself. **At a first reading, though, you may prefer to leave this out and go straight on to section 10.3.**

The relationship between vapour pressure and temperature is obtained in the following way. Remember that in Topic 3 we defined entropy change, ΔS, as $k\Delta \ln W$, where k is the Boltzmann constant.

In Topic 3 we saw that if one mole of a gas at constant temperature expands from volume V_1 to volume V_2, the number of arrangements of molecules is multiplied by

$(V_2/V_1)^L$ (where L is the Avogadro constant)

Writing W_1 and W_2 for the number of ways of arranging molecules

$$W_2/W_1 = (V_2/V_1)^L$$

Taking logarithms:

$$\ln(W_2/W_1) = L \ln(V_2/V_1)$$

In the present case we need a relationship involving pressures.
For ideal gases

$$V_2/V_1 = p_1/p_2$$

so

$$\ln(W_2/W_1) = L \ln(V_2/V_1) = -L \ln(p_2/p_1)$$

So the change in the logarithm of the number of arrangements is $-L$ times the change in the logarithm of the pressure, or

$$\Delta \ln W = -L \Delta \ln p$$

It follows that

$$\Delta S (= k \Delta \ln W) = -Lk \Delta \ln p$$

This is the relationship involving pressures that we need.
It tells us that the change in entropy, ΔS, of a gas, which takes place because of a change in pressure, Δp, is given by the equation

$$\Delta S = -Lk \Delta \ln p$$

Now let us suppose that $\Delta S_{evap}^{\ominus}$ is the entropy change on vaporization at standard pressure $p^{\ominus}$, and ΔS_{evap} is the entropy change at another pressure p. Then we may write

$$\Delta S_{evap} = \Delta S_{evap}^{\ominus} - Lk \ln p/p^{\ominus}$$

The equation for equilibrium, mentioned in section 10.1, is

$$\Delta S_{evap} - \Delta H/T = 0$$

As ΔS_{evap} will be altered by changing the pressure, so the value of T at which equilibrium exists will be altered. The equation linking p and T is given by combining the last two equations:

$$\Delta H/T = \Delta S^{\ominus}_{evap} - Lk \ln p/p^{\ominus}$$

This can be rearranged to give

$$Lk \ln p/p^{\ominus} = -\Delta H/T + \Delta S^{\ominus}_{evap}$$

Thus a graph of $Lk \ln p/p^{\ominus}$ against $1/T$ should be a straight line, with gradient $-\Delta H$ (see figure 10.4).

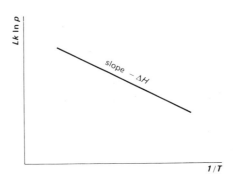

Figure 10.4

The meaning of this graph is as follows. As p falls, the expansion entropy rises as $-Lk \ln p$, and so can afford a bigger fall in the entropy of the surroundings. The energy ΔH can therefore be absorbed from the surroundings at a lower temperature T. The falling graph shows one entropy change 'paying' for another.

Another way to write the relationship is

$$Lk \ln p = \Delta S^{\ominus}_{evap} - Lk \ln p^{\ominus} - \Delta H/T$$

The first two terms on the right are constant, so we get

$$Lk \ln p = \text{constant} - \Delta H/T$$

or

$$p = (\text{another constant}) \exp(-\Delta H/LkT)$$

The relationship between p and T is sketched in the graph below.

The results emphasize, now quantitatively, what has been seen before: if you want a reaction to happen, reduce the partial pressure of the product.

Example   As an illustration of the use of this relationship, let us apply it to solve the following problem.

At an altitude of 5000 feet (just over 1500 metres) the atmospheric pressure is about 650 mmHg. At what temperature would you expect water to boil at this altitude? (ΔH_b for water is 40.6 kJ mol^{-1}.)

Substituting the equation $Lk \ln p = \text{constant} - \Delta H/T$ we have, for 760 mmHg, and a boiling point of 373 K,

$$6.02 \times 10^{23} \times 1.38 \times 10^{-23} \times \ln 760 = \text{constant} - \frac{40\,600}{373} \qquad \textbf{1}$$

and for 650 mmHg, and a boiling point of TK,

$$6.02 \times 10^{23} \times 1.38 \times 10^{-23} \times \ln 650 = \text{constant} - \frac{40\,600}{T} \qquad \textbf{2}$$

Subtracting equation **2** from equation **1** we have

$$6.02 \times 10^{23} \times 1.38 \times 10^{-23} \left(\ln \frac{760}{650} \right) = -\frac{40\,600}{373} + \frac{40\,600}{T}$$

from which

$$T = 369 \text{ K } (96\,°\text{C})$$

So, at an altitude of 5000 feet, water boils at 96 °C.

Make several calculations for different pressures, and compare your results with those that you obtained in experiment 10.2.

10.3
RAOULT'S LAW

In the last section we considered the effect of temperature on the vapour pressure of water, that is, how temperature affects the pressure of water vapour in equilibrium with liquid water. We then saw how entropy considerations could be used to explain our observations.

We shall now carry this investigation a step further. Suppose an involatile solute is dissolved in the water; how will this affect the vapour pressure of the water?

This problem was studied experimentally in 1886 by the French chemist François Marie Raoult, and he found that the vapour pressure above a solution of an involatile solute was *less* than that above the pure solvent at the same temperature. He quantified his results in the form of a law, now known as Raoult's Law.

There are several ways of stating Raoult's Law. One form of it is as follows:

The vapour pressure of a solvent above a solution is proportional to the mole fraction of the solvent present.

The constant of proportionality in this relation is, in fact, the vapour pressure of the pure solvent at the same temperature. So we may say:

The vapour pressure of a solvent above a solution is the vapour pressure of the pure solvent at the same temperature multiplied by the mole fraction of the solvent present.

Expressed in symbols this is

$$p = p^\circ \times \frac{N}{n + N}$$

where

p is the vapour pressure of the solution
p° is the vapour pressure of the pure solvent
N is the number of moles of solvent
n is the number of moles of solute

Example The vapour pressure of water at $100\,^\circ C$ is $760\,mmHg$. What is the vapour pressure of water above a solution of urea containing 6 g of urea in 100 g of water at the same temperature?

The relative molecular mass of urea, CON_2H_4, is 60. The amount of urea present is $6/60 = 0.10$ mole. The amount of water present is $100/18 = 5.56$ moles. The vapour pressure of water above this solution at $100\,°C$ is p where

$$p = 760\left(\frac{5.56}{0.10 + 5.56}\right) = 747\,mmHg$$

Raoult's Law and entropy

Let us now see how our work on entropy can help us to understand why the vapour pressure of the solvent is lowered in this way. This time we shall examine the situation, which is quite a complicated one, in a purely qualitative manner, and not attempt any mathematics.

The molecules of the solute are mixed with the water molecules in the solution, and not arranged regularly in a solid structure. This fact might be considered to increase the number of ways in which the various particles and their energy can be arranged in the liquid. The entropy of the solution might therefore be considered to be higher than that of the pure solvent. However, the presence of the solute, which is involatile, will not alter the entropy of the vapour in equilibrium with the liquid, because there is none of the solute in the vapour.

If we compare pure water with the solution of the involatile solute in water at the same temperature, then the entropy decrease of the surroundings when some water evaporates, $\Delta H/T$, is going to be much the same for both (provided that the solution is dilute).

However, if the entropy of the solution is greater than that of the water alone, the entropy increase of the system will be less for the solution than it will be for the water alone if both form vapour at the same pressure.

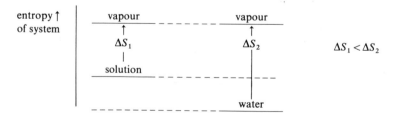

There is only one way in which a solution of higher entropy than that of pure water can produce the required entropy increase to match the entropy decrease of the surroundings. This is to produce vapour at a lower pressure than that in equilibrium with the water alone.

These ideas therefore provide us with a model to help us understand why the vapour pressure is lowered in this way.

How does Raoult's Law apply to mixtures of liquids?

When two liquids are mixed together, each acts like a solvent dissolving the other. This means that the vapour pressure of each is reduced.

Suppose we have two liquids that mix together in all proportions, and suppose Raoult's Law is obeyed over the whole range of possible proportions. The vapour pressure of A will be at a maximum in pure A, and will fall uniformly as the mole fraction is reduced, until it reaches zero. This is shown by the line marked A in figure 10.5. The vapour pressure of B will vary in a similar manner, as shown by line B in the figure.

The total vapour pressure given by both liquids thus varies with the composition of the mixture as shown by the line marked 'total' in figure 10.5.

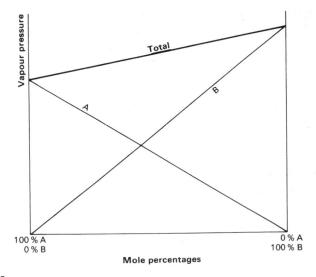

Figure 10.5

The contribution of each liquid to the total vapour pressure above a mixture can be calculated from the Raoult's Law relationship, which can now be written

$$p_A = p_A^\circ \times \frac{n_A}{n_A + n_B}$$

where

p_A is the vapour pressure of liquid A above the mixture
p_A° is the vapour pressure of pure A
n_A is the number of moles of A in the mixture
n_B is the number of moles of B in the mixture

Example At 60 °C the vapour pressure of pure benzene is 385 mmHg, and that of methylbenzene (toluene) is 139 mmHg. What is the total vapour pressure above a mixture of 60 g of benzene and 40 g of methylbenzene at 60 °C?

The relative molecular mass of benzene, C_6H_6, is 78, and that of methylbenzene, C_6H_5—CH_3, is 92. The mixture therefore contains 60/78 = 0.77 mole benzene, and 40/92 = 0.43 mole methylbenzene.

The vapour pressure of benzene, p_B, is given by

$$p_B = 385 \times \frac{0.77}{0.77 + 0.43} = 247 \text{ mmHg}$$

The vapour pressure of methylbenzene, p_m, is given by

$$p_m = 139 \times \frac{0.43}{0.77 + 0.43} = 50 \text{ mm Hg}$$

The total vapour pressure above the mixture is therefore

247 + 50 = 297 mmHg

Figure 10.5 is, of course, concerned with vapour pressures at a constant temperature. The vapour pressure above a liquid increases with increasing temperature until it becomes equal to the atmospheric pressure, when the liquid boils. At a given atmospheric pressure, liquids having a high vapour pressure at a particular temperature are likely to have a lower boiling point than liquids having a low vapour pressure. The boiling points of liquid mixtures at a given pressure must therefore vary with composition, because their vapour pressures at a given temperature vary in this way. In experiment 10.3a you will investigate just how boiling points of a liquid mixture vary with composition, for a pair of liquids that obey Raoult's Law quite closely.

EXPERIMENT 10.3a
How does the boiling point of a mixture of propan-1-ol and propan-2-ol vary with its composition?

The reflux apparatus you will use is illustrated in figure 10.6. The thermometer should dip into the liquid mixture, but not touch the walls of the flask.

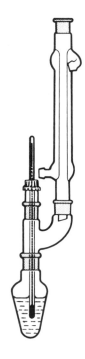

Figure 10.6

Warning: the purpose of the reflux condenser is to prevent loss of vapour and *reduce the risk of fire.* Remember, when handling flammable liquids and especially when adding samples to the apparatus, to *turn the Bunsen burner off.*

Before you begin to determine the various boiling points, make a copy of the following table in your notebook. For each experiment you will have a different proportion of the two liquids. You will be measuring the liquids by volume; the corresponding mole fractions have already been worked out for you, and are given in the table.

Experiment number	Volume/cm^3		Mole fraction		Boiling point/$^\circ$C
	Propan-1-ol	**Propan-2-ol**	**Propan-1-ol**	**Propan-2-ol**	
1	10	0	1.000	0.000	
2	10	2	0.835	0.165	
3	10	4	0.717	0.283	
4	10	6	0.628	0.372	
5	10	8	0.559	0.441	
6	10	10	0.503	0.497	
7	0	10	0.000	1.000	
8	2	10	0.168	0.832	
9	4	10	0.288	0.712	
10	6	10	0.378	0.622	
11	8	10	0.448	0.552	

Procedure

Experiment 1 Measure 10.0 cm^3 of propan-1-ol directly into the flask, add two or three anti-bumping granules, and heat the flask with a small Bunsen burner flame until the liquid is boiling gently. Record the temperature, which is the boiling point of pure propan-1-ol, in the 'boiling point' column of your table, on the line for experiment number 1.

Experiment 2 Measure 2.0 cm^3 of propan-2-ol from a burette into a test-tube, and after turning the Bunsen burner off and allowing the apparatus to cool briefly, add it to the contents of the flask by pouring it down the condenser. Reheat the mixture until it is boiling gently and record the new boiling point in the table.

Experiments 3 to 6 Take further readings after additions of 2.0-cm^3 portions until a total of 10.0 cm^3 of the propan-2-ol has been added, recording each result in your table.

Experiments 7 to 11 Empty out the flask, disposing of the contents as instructed; the liquids should not be poured straight down the sink. Then repeat experiments 1 to 5, starting with 10.0 cm³ of propan-2-ol in the flask, and make 2.0-cm³ additions of propan-1-ol.

Plot a graph of boiling point against molar composition. Is there a linear relationship between them?

It seems likely that the boiling points at constant pressure for mixtures of liquids are inversely proportional to their vapour pressures at constant temperature, though the relationship may not be an exact one.

We should not be surprised, therefore, if a boiling point–composition curve for a pair of liquids that obey Raoult's Law is roughly linear, with a slope in the opposite direction to that of the vapour pressure–composition curve. This is the case with the liquids used in experiment 10.3a. A pair of liquids which obey Raoult's Law is said to form an *ideal solution*.

There are, however, many pairs of liquids that do not obey Raoult's Law particularly closely, forming what are called *non-ideal solutions*. What do you suppose their boiling point–composition curves look like?

Deviations from Raoult's Law

We can distinguish two types of deviation from Raoult's Law. If the vapour pressure of a mixture of liquids is greater than that predicted for an 'ideal' solution, this is known as a *positive deviation* from Raoult's Law. If the vapour pressure is less than that predicted, it is known as a *negative deviation* from Raoult's Law. Typical vapour pressure–composition curves for these deviations are shown in figure 10.7.

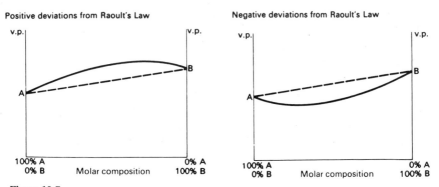

Figure 10.7
Vapour pressure–composition curves for pairs of liquids that deviate from Raoult's Law.

What effect do you think each of these deviations will have on the boiling point–composition curves? You can test your predictions in experiment 10.3b, in which you will be using pairs of liquids which do not obey Raoult's Law very closely.

EXPERIMENT 10.3b
How does the boiling point of a mixture of liquids that do not obey Raoult's Law closely vary with its composition?

Repeat experiment 10.3a, this time using either

1 cyclohexane and ethanol, or
2 trichloromethane and ethyl ethanoate.

WARNING: trichloromethane should only be handled in an efficient fume cupboard. Prolonged exposure to the substance may cause cancer. Cyclohexane, ethanol, and ethyl ethanoate are all highly flammable.

Copy the following tables into your notebook, entering your results as before, and using those of another group for the pair of liquids you have not tested.

Experiment number	Volume/cm^3		Mole fraction		Boiling point/$^\circ$C
	Ethanol	Cyclohexane	Ethanol	Cyclohexane	
1	10	0	1.000	0.000	
2	10	2	0.902	0.098	
3	10	4	0.822	0.178	
4	10	6	0.755	0.245	
5	10	8	0.698	0.302	
6	10	10	0.649	0.351	
7	0	10	0.000	1.000	
8	2	10	0.270	0.730	
9	4	10	0.425	0.575	
10	6	10	0.526	0.474	
11	8	10	0.597	0.403	

Experiment number	Volume/cm^3		Mole fraction		Boiling point/°C
	Trichloro-methane	Ethyl ethanoate	Trichloro-methane	Ethyl ethanoate	
1	10	0	1.000	0.000	
2	10	2	0.858	0.142	
3	10	4	0.752	0.248	
4	10	6	0.669	0.331	
5	10	8	0.602	0.398	
6	10	10	0.548	0.452	
7	0	10	0.000	1.000	
8	2	10	0.195	0.805	
9	4	10	0.326	0.674	
10	6	10	0.421	0.579	
11	8	10	0.492	0.508	

Plot graphs of boiling point against molar composition for each pair of liquids.

The results that you obtain should be like the graphs shown in figure 10.8. Classify the cyclohexane–ethanol system and the trichloromethane–ethyl ethanoate system into their appropriate categories.

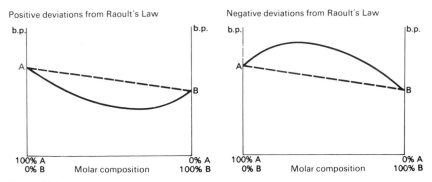

Figure 10.8
Boiling point–composition curves for pairs of liquids that deviate from Raoult's Law.

Do the graphs in figure 10.8 agree with what you expected after you had studied figure 10.7?

What causes the deviations from Raoult's Law?

It is possible that some pairs of liquids do not obey Raoult's Law because their molecules interact with each other. If this is the case, we would expect that some enthalpy change would take place. Whether or not this happens is investigated in the next experiment.

EXPERIMENT 10.3c
Do enthalpy changes occur when non-ideal solutions are made?

In this experiment you should investigate the pair of liquids that you used in experiment 10.3b, either

1 cyclohexane and ethanol, or
2 trichloromethane and ethyl ethanoate.

WARNING: trichloromethane should only be handled in an efficient fume cupboard. Prolonged exposure to the substance may cause cancer. Remember also that three of these liquids are flammable, and all of them are volatile. Make sure that no Bunsen burners are alight, and see that the laboratory is well ventilated. When the experiment is over, the liquids must be disposed of as you are instructed; they must not be poured away down the sink.

Procedure

1 Put each pure liquid in a burette. Put a boiling-tube in a beaker and surround the tube with a heat-insulating material, such as cottonwool. Keep a second boiling-tube available in a test-tube rack.

2 Run $18.0\,cm^3$ of one liquid into the insulated boiling-tube, and then run $2.0\,cm^3$ of the other liquid into the second boiling-tube. Record the temperature of each liquid.

3 Tip the liquid in the free tube into the insulated tube; stir gently with the thermometer, and record the temperature.

4 Repeat for $15.0\,cm^3$ of the liquid and $5.0\,cm^3$ of the second, and continue the process until a complete range of values has been obtained. Use a total of $20.0\,cm^3$ of mixture on each occasion.

Volume/cm^3						
Liquid A	18	15	12	8	5	2
Liquid B	2	5	8	12	15	18

Plot a graph of temperature change against volume composition.

How does this graph compare with the graph of boiling point against molar composition?

How many moles of A and how many moles of B are there at any point of particular interest in the graph?

What interpretation in terms of bonding would you put on
1 an endothermic change
2 an exothermic change?

It is apparent from the experimental work on non-ideal mixtures and their enthalpy change of mixing that intermolecular bonding occurs when substances such as trichloromethane and ethyl ethanoate are allowed to mix. Further qualitative tests can now be carried out to find out which atoms are involved in this bonding.

Suppose we measure the temperature change that takes place when 10-cm^3 portions of the following liquids are mixed:

dichloromethane and ethyl ethanoate
trichloromethane and ethyl ethanoate
tetrachloromethane and ethyl ethanoate

Because of the hazards involved in the storage and handling of these substances, especially tetrachloromethane, you will probably not be able to do this yourself. Typical results, are, however:

for dichloromethane and ethyl ethanoate, a rise of 5.8 °C
for trichloromethane and ethyl ethanoate, a rise of 10.4 °C
for tetrachloromethane and ethyl ethanoate, a fall of 1.0 °C

Using these results as evidence, would you suggest that the substituted methanes bond to ethyl ethanoate by means of a hydrogen or a chlorine atom?

Would you suggest that trichloromethane bonds to a methyl group or an oxygen atom of ethyl ethanoate?

Experiments such as the one above show us that for very many examples of intermolecular bonding a hydrogen atom is necessary, and that the bonding takes place by means of this atom. Such bonds are known as *hydrogen bonds*.

Hydrogen bonds occur between atoms of hydrogen and the small, strongly electronegative atoms of the elements nitrogen, oxygen, and fluorine; and these latter atoms must be in molecular situations in which they have available at least one pair of non-bonded electrons. Because of the small size of the hydrogen atom, and the comparatively small size of the other atoms involved, the two atoms are able to approach one another closely enough for the forces of attraction between them to reach nearly one-tenth of the strength of a typical covalent bond.

Hydrogen bonds are symbolized by three dots, thus:

$$H_3N \cdots H—OH, \quad H—F \cdots H—F$$

Bond energies of typical hydrogen bonds are around $25 \, kJ \, mol^{-1}$. Those of typical covalent bonds are about $300–400 \, kJ \, mol^{-1}$.

10.4
THE INTERPRETATION OF PROPERTIES IN TERMS OF HYDROGEN-BOND FORMATION

A number of the properties of compounds can be interpreted as being caused by the existence of hydrogen bonding. We shall now consider some examples of these properties, beginning with the failure of certain mixtures of liquids to obey Raoult's Law.

1 Non-ideal behaviour

The vapour pressure of a liquid is a measure of the tendency of the molecules within it to escape. Thus, if one liquid is added to another and a solution is formed whose boiling point is higher than expected on the basis of ideal behaviour, some phenomenon must be reducing the escaping tendency. There must be a greater measure of intermolecular attraction than before. In instances where marked deviation from the ideal occurs, the attraction is strong enough to be classed as a weak bond. In the case of trichloromethane and ethyl ethanoate, this weak bond is a hydrogen bond.

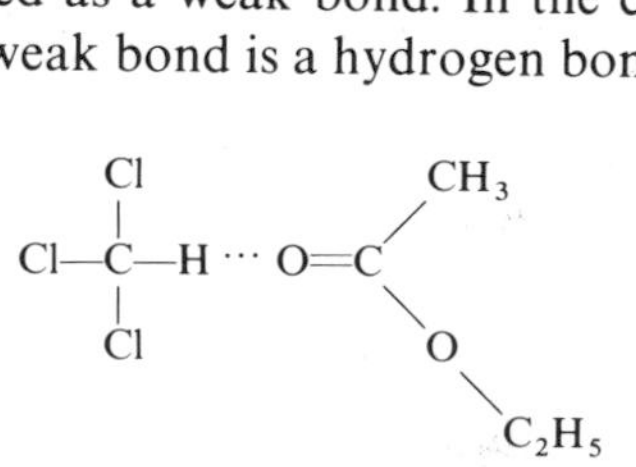

If a solution is formed whose boiling point is lower than expected on the basis of ideal behaviour, then some phenomenon is increasing the escaping tendency. It must be reducing the intermolecular attractions existing in the pure liquids. Thus in a mixture of ethanol and cyclohexane the intermolecular attraction must be less than in pure ethanol or pure cyclohexane. This may be interpreted by visualizing molecules of one liquid becoming so numerous in the mixture that they interfere with intermolecular attraction between the molecules of the other liquid.

In pure ethanol a very high proportion of the molecules are hydrogen-bonded. For instance

$$C_2H_5-O \diagdown H \qquad \begin{array}{c} H \\ \diagdown \\ O-C_2H_5 \end{array}$$

$$O-H \cdots O$$
$$\diagup \qquad \diagdown$$
$$C_2H_5 \qquad C_2H_5$$

When cyclohexane is added, the cyclohexane molecules get in between the molecules of ethanol, breaking up the weak hydrogen bonds, and markedly reducing the previous intermolecular attraction. The vapour pressure thus rises and the boiling point falls more than expected in an ideal system.

Further evidence for the presence of hydrogen bonds in alcohols, and the effects on such bonds of the addition of non-polar molecules, is obtained from a study of infra-red spectra.

In ideal mixtures such as propan-1-ol and propan-2-ol, the attractions within the pure liquids are so nearly the same as those in mixtures that Raoult's Law is closely obeyed.

Many negative deviations from Raoult's Law may be attributed to hydrogen bond formation in the mixture, and a range of positive deviations may be attributed to the breaking of hydrogen bonds in one component by the interfering action of the other. A further example of a pair of compounds showing a negative deviation is propanone and phenylamine (aniline); a further example of a positive deviation is given by a mixture of ethanol and benzene.

It must be remembered, however, that other weak intermolecular forces such as those discussed later in this Topic may also be responsible in part for observed deviations from Raoult's Law.

2 Anomalous properties of certain hydrides

The boiling points of the Group IV hydrides (figure 10.9) decrease with decreasing relative molecular mass from tin to carbon. But in Groups V, VI, and VII this is not so, for extrapolation of the general trend in each group would give a much lower boiling point for the first member than actually occurs. Graphs of melting points, and of enthalpy changes of vaporization, show similar patterns.

Examine the data in table 10.2. What do you suppose is the effect of hydrogen bonding on the enthalpy changes of fusion, $\Delta H_m^{\ominus}$, enthalpy changes of vaporization, $\Delta H_b^{\ominus}$, and melting points of the hydrides of the p-block elements?

Substance	Melting point /°C	Boiling point /°C	$\Delta H_m^{\ominus}$ /kJ mol^{-1}	$\Delta H_b^{\ominus}$ /kJ mol^{-1}
CH_4	-182	-161	0.92	8.20
SiH_4	-185	-112	0.67	12.1
GeH_4	-165	-88	0.84	14.1
SnH_4	-150	-52	—	18.4
PbH_4	—	-13	—	—
NH_3	-78	-33	5.65	23.4
PH_3	-133	-88	1.13	14.6
AsH_3	-116	-55	2.34	17.5
SbH_3	88	-17	—	—
BiH_3	—	$+22$	—	—
OH_2	0	$+100$	6.02	40.7
SH_2	-85	-61	2.39	18.7
SeH_2	-66	-41	2.51	19.3
TeH_2	-49	-2	—	23.2
HF	-83	$+20$	4.56	7.74
HCl	-115	-85	2.01	16.2
HBr	-88	-67	2.43	17.6
HI	-51	-35	2.89	19.8

Table 10.2
Data on the hydrides of the p-block elements.

3 The dimerization of organic acids

When ethanoic acid, CH_3CO_2H ($M = 60$), and benzoic acid, $C_6H_5CO_2H$ ($M = 122$) are dissolved in benzene, it is found that the solute particles have relative molecular masses of nearly 120 and 244 respectively. To account for these observations it is suggested that the molecules must exist in solution as dimers, held together by hydrogen bonds:

When ethanoic acid is dissolved in water or alcohols the solute particles have the expected relative molecular mass of 60. Why do dimers not form in these cases?

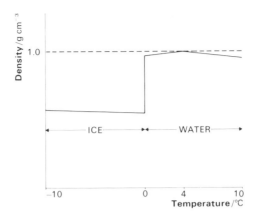

Figure 10.12
Density changes during the conversion of ice to water (density not to scale).

Further evidence for the existence of dimers of organic acids has been obtained from X-ray diffraction studies of crystals of sorbic acid, $CH_3(CH{=}CH)_2CO_2H$ (see figure 10.13).

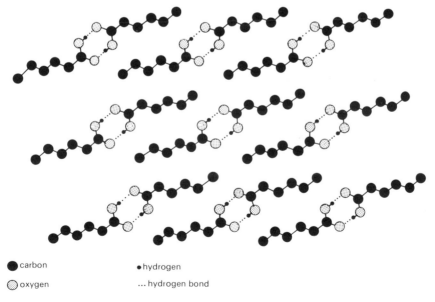

● carbon   ● hydrogen

◎ oxygen   ... hydrogen bond

Figure 10.13
Sorbic acid structures (hydrogen atoms of the alkyl groups omitted).

10.5
OTHER INTERMOLECULAR FORCES

Earlier in the Topic, we studied the behaviour of tetrachloromethane and ethyl ethanoate on mixing, and observed a slight temperature drop. Does this suggest an increase or decrease in intermolecular bonding when the pure liquids are mixed? What properties of the mixture might be affected by changes in intermolecular bonding?

The tetrachloromethane molecule contains no hydrogen atom to form a hydrogen bond to a molecule of ethyl ethanoate. It is also a symmetrical molecule. Is the ethyl ethanoate molecule symmetrical in relation to distribution of charge due to its electrons? What previous experimental tests have indicated the charge distribution in ethyl ethanoate?

Can the distribution of charge in the ethyl ethanoate molecule be used to explain the enthalpy change of mixing with tetrachloromethane?

Van der Waals forces

The noble gases provide evidence for the existence of cohesive forces between molecules. Helium, which does not form normal bonds and has symmetrical atoms, condenses to a liquid and ultimately freezes to a solid at very low temperatures. Energy is evolved in this process, showing that cohesive forces are operating.

The energy of sublimation of solid helium is only

$$0.105 \, \text{kJ mol}^{-1} \qquad (\Delta H_m + \Delta H_b)$$

This should be compared with the energy of sublimation of ice, $46.9 \, \text{kJ mol}^{-1}$, which is used to overcome hydrogen bonding, and the dissociation energy of the oxygen molecule, $494 \, \text{kJ mol}^{-1}$, required to break the two covalent bonds.

The weak forces of attraction, independent of normal bonding forces, which are found to exist between atoms and molecules in the solid, liquid, and gas states are known as *van der Waals forces*.

Van der Waals forces are considered to be due to continually changing dipole-induced dipole interactions between atoms.

These dipoles are thought to arise because the electron charge-cloud in an atom is in continual motion. In the turmoil it frequently happens that rather more of the charge-cloud is on one side of the atom than on the other. This means that the centres of positive and negative charge do not coincide, and a fluctuating dipole is set up. This dipole induces a dipole in neighbouring atoms. The sign of the induced dipole is opposite to that of the dipole producing it, and consequently a force of attraction results. These flickering atomic dipoles and induced dipoles produce a cohesive force between neighbouring atoms and molecules.

The greater the number of electrons in an atom, the greater will be the fluctuation in the asymmetry of the electron charge-cloud and the greater will be the van der Waals attraction set up. The rise in boiling point down Group VII – fluorine, chlorine, bromine, and iodine – is due to the increase in electrons present in the atoms and the consequent increase in van der Waals attractions, rather than to the increase in the mass of the atoms.

The increase in boiling point up the homologous series of alkanes is due to the increased number of electrons in the molecules and the increased total van der Waals attractions rather than to the increase in mass of the molecules. Similarly the difference in boiling point between isomers can be explained in terms of van der Waals attractions.

The structures of pentane and 2,2-dimethylpropane, two isomers of molecular formula C_5H_{12}, are shown in figures 10.15 and 10.16. Their boiling points are 36 °C and 9 °C respectively.

Build space-filling models of each of these structures, and compare the shapes of the molecules.

Can you suggest a reason for the difference in the boiling points of these isomers?

pentane
M = 72
b.p. = 36 °C

2,2-dimethylpropane
M = 72
b.p. = 9 °C

Figures 10.15 and 10.16
Two isomers of molecular formula C_5H_{12}.

In the case of pentane, the linear molecules can line up beside each other and the van der Waals forces are likely to be comparatively strong, as they can act over the whole of the molecule. In the case of 2,2-dimethylpropane, the spherical molecules can only become close to one another at one point, so the van der Waals forces are likely to be comparatively weak. The isomer with the linear molecules thus has the higher boiling point.

Van der Waals radii

The normal bonding forces in molecules are concentrated within the molecules themselves; they are intramolecular. In crystals individual molecules are held to each other by van der Waals forces. Examples are iodine, solid carbon dioxide, and naphthalene. (As the forces are weak, the melting points of molecular crystals tend to be low. If permanent dipoles, or hydrogen bonding, are present as well, then the melting point will be higher.)

In molecular crystals the van der Waals forces draw molecules together until their electron charge-clouds repel each other to the extent of balancing the attraction. Thus for argon the atoms are drawn together until the atomic nuclei have a separation of about 0.4 nm (figure 10.17).

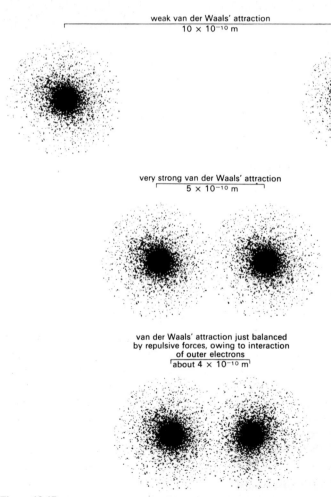

weak van der Waals' attraction
10×10^{-10} m

very strong van der Waals' attraction
5×10^{-10} m

van der Waals' attraction just balanced
by repulsive forces, owing to interaction
of outer electrons
about 4×10^{-10} m

Figure 10.17
Van der Waals attraction between argon atoms.

The atomic distances within simple molecules and between simple molecules are not the same. The covalent radius is one half of the distance between two atoms in the same molecule. The van der Waals radius is one half of the distance between the nuclei of two atoms in adjacent molecules (see figure 10.18). The relative values of these radii can be seen in table 10.3.

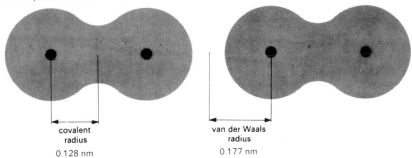

covalent radius
0.128 nm

van der Waals radius
0.177 nm

Figure 10.18
Covalent radius and van der Waals radius for molecules of I_2 in an iodine crystal.

Atom	Covalent radius/nm	Van der Waals radius/nm
H	0.037	0.12
N	0.075	0.15
O	0.073	0.15
P	0.110	0.19
S	0.102	0.18

Table 10.3
Comparative values for covalent and van der Waals radii.

The importance of van der Waals forces

Although the van der Waals forces between individual atoms, as in argon, give only a small bonding energy, the total van der Waals bonding energy can be significant in large molecules with many contacts between atoms. For example, a well crystallized sample of polythene (the high density form) has a tensile strength of $30 \times 10^6 \, N \, m^{-2}$; but polythene of poorer crystallinity, that is, with less orderly packing (the low density form), has reduced van der Waals bonding energy and a tensile strength of $10 \times 10^6 \, N \, m^{-2}$.

Van der Waals forces can also make an important contribution to the structure of globular proteins. Molecules such as proteins tend to form aggregates in aqueous solution because of their hydrocarbon side-chains. These aggregates can be dispersed by the addition of urea, which is known to break weak bonds. Thus, on the addition of urea, haemoglobin will break into two different sub-units. If the urea is then removed, the sub-units will associate to reform haemoglobin molecules. The process is known to be exact because the reformed haemoglobin is physiologically active.

SUMMARY

At the end of this Topic you should:

1 be able to calculate the entropy change of vaporization of a liquid, given the enthalpy change of vaporization and the boiling point;

2 know Trouton's Rule, understand why the rule holds, and be able to offer likely reasons why certain substances are exceptions to the rule;

3 know how to investigate the change of vapour pressure with temperature, for a liquid;

4 be able to use the relationship between vapour pressure and temperature, $Lk \ln p = \text{constant} - \Delta H/T$;

5 know Raoult's Law and be able to use it to work out the vapour pressure of solutions of involatile solutes and of mixtures of liquids;

6 know how to investigate the variation of the boiling point of mixtures of liquids, with the composition of the liquids;

7 understand what is meant by an 'ideal solution', and by positive and negative deviations from Raoult's Law;

8 be able to explain deviations from Raoult's Law in terms of intermolecular forces;

9 understand what is meant by a hydrogen bond and be able to discuss examples of hydrogen bonding;

10 understand what is meant by van der Waals forces, and be able to discuss examples of this type of intermolecular attraction.

PROBLEMS

1 For a number of liquids, table 10.1 gives the enthalpy changes of vaporization and the boiling points. Using this table, find the entropy changes of vaporization for

a tetrachloromethane
b octane
c ethanol

Explain why similar values are obtained for tetrachloromethane and octane. Why do you think that ethanol has a higher value?

2 The vapour pressure of water at 100 °C is 760 mmHg. What is the vapour pressure of water above the following solutions, at the same temperature?

a 15 g glucose, $C_6H_{12}O_6$, in 100 g water
b 4 g urea, CON_2H_4, in 162 g water.

3 A solution of 10.6 g glucose in 100 g water had a vapour pressure of 752 mmHg at 100 °C. The vapour pressure of water at 100 °C is 760 mmHg. What value do these figures suggest for the relative molecular mass of glucose?

4 Figure 10.19 shows the vapour pressure at constant temperature of mixtures, of various composition, of two liquids X and Y. State which of the following pairs of liquids are most likely to behave like X and Y, and give reasons for your choice.

A propan-1-ol and propan-2-ol
B propan-1-ol and methylbenzene (toluene)
C benzene and methylbenzene
D benzene and heptane
E propan-2-ol and ethanol

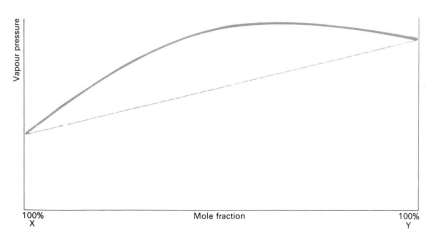

Figure 10.19

5 Arrange the following in the order you should expect for their boiling points, putting the one with the highest boiling point first.

A $CH_3CH_2CH_2CH_3$ C $CH_3CH_2CH_2CH_2Cl$

B $CH_3-\underset{\underset{CH_3}{|}}{\overset{\overset{CH_3}{|}}{C}}-H$ D $CH_3-\underset{\underset{CH_3}{|}}{\overset{\overset{CH_3}{|}}{C}}-Cl$

Give reasons for your answer.

6 The vapour pressures of octane and 2-methylheptane at 30 °C are 19.00 mmHg and 27.4 mmHg respectively.

a Calculate the vapour pressure of a mixture of the two liquids containing the mole fraction 0.4 of octane at 30 °C.
b What assumption about the mixture did you need to make before you could calculate its vapour pressure? On what evidence did you decide that the assumption was reasonable?

7 Classify the following mixtures of liquids into:
A those likely to obey Raoult's Law
B those likely to show a positive deviation from Raoult's Law
C those likely to show a negative deviation from Raoult's Law

a propanone and butanone
b 1,1,2,2-tetrachloroethane and ethanal
c methylbenzene and dimethylbenzene
d trichloromethane and ethoxypropane
e benzyl alcohol and benzene
   State briefly the reasons for your answers.

8 Arrange each of the following groups of liquids in the order you would expect for their boiling points, putting the liquid with the highest boiling point first.
a helium, neon, argon
b propane, butane, pentane
c hydrogen fluoride, hydrogen chloride, sodium chloride
d hydrazine, silicoethane, diborane
e benzoic acid, 4-hydroxybenzoic acid, benzene
   Give reasons for your answers to d and e.

Alcohols, carbonyls, and carbohydrates

In this second Topic on the organic compounds of carbon we are going to look at two new functional groups both of which contain oxygen. These are

the hydroxyl group, —OH, as in ethanol, and

the carbonyl group, $\diagdown$C$=$O, as in propanone, and propanal.

a Displayed formulae

$$
\begin{array}{ccc}
\text{H} \ \ \text{H} & \text{H} \ \ \text{O} \ \ \text{H} & \text{H} \ \ \text{H} \ \ \text{O} \\
| \ \ \ | & | \ \ \| \ \ | & | \ \ | \ \ \\
\text{H--C--C--O--H} & \text{H--C--C--C--H} & \text{H--C--C--C} \\
| \ \ \ | & | \ \ \ \ \ | & | \ \ | \ \ \diagdown \\
\text{H} \ \ \text{H} & \text{H} \ \ \ \ \ \text{H} & \text{H} \ \ \text{H} \ \ \ \text{H}
\end{array}
$$

b Structural formulae

$$CH_3—CH_2—OH \qquad CH_3—CO—CH_3 \qquad CH_3—CH_2—CHO$$
$$\text{ethanol} \qquad\qquad \text{propanone} \qquad\qquad \text{propanal}$$

We shall also consider the properties of the very important group of natural products known as carbohydrates. Carbohydrates are considered in this Topic because they contain both the functional groups we are discussing.

11.1
ALCOHOLS

Alcohols have a hydroxyl functional group (—OH) attached to a saturated carbon atom. They are named from the parent alkane, with the terminal 'e' replaced by the ending 'ol'.

$$
\begin{array}{llll}
& CH_4 & \text{methane} & \qquad CH_3—OH \quad \text{methanol} \\
CH_3—CH_3 & & \text{ethane} & \qquad CH_3—CH_2—OH \quad \text{ethanol} \\
CH_3—CH_2—CH_3 & & \text{propane} & \qquad CH_3—CH_2—CH_2—OH \quad \text{propan-1-ol}
\end{array}
$$

The rules for naming the isomers are similar to those outlined for hydrocarbons and halogenoalkanes. If necessary, you should reread the section entitled

'Some rules for naming organic compounds' in Topic 9.

If the —OH group is attached to a carbon atom which is itself attached directly to only one other carbon atom, the compound is known as a *primary alcohol*. If the —OH group is attached to a carbon atom which is attached directly to two other carbon atoms, the compound is a *secondary alcohol*, and if to three other carbon atoms, it is a *tertiary alcohol*. Make sure that you understand this naming by looking carefully at the following structural formulae, which are all isomers of $C_4H_{10}O$.

$$CH_3—CH_2—CH_2—CH_2—OH \quad \text{butan-1-ol, a primary alcohol}$$

$$CH_3—CH_2—\underset{\underset{OH}{|}}{CH}—CH_3 \qquad\qquad \text{butan-2-ol, a secondary alcohol}$$

$$CH_3—\underset{\underset{OH}{|}}{\overset{\overset{CH_3}{|}}{C}}—CH_3 \qquad\qquad \text{2-methylpropan-2-ol, a tertiary alcohol}$$

If alcohols contain more than one hydroxyl group they are known as diols or triols, etc., after the number of hydroxyl groups they contain.

$$\underset{\underset{OH}{|}}{CH_2}—\underset{\underset{OH}{|}}{CH_2} \qquad \text{ethane-1,2-diol (glycol)}$$

$$\underset{\underset{OH}{|}}{CH_2}—\underset{\underset{OH}{|}}{CH}—\underset{\underset{OH}{|}}{CH_2} \qquad \text{propane-1,2,3-triol (glycerol)}$$

The infra-red spectra of alcohols have a set of peaks characteristic of their structures, the most important being the O—H stretching wavenumber at $3600\,\text{cm}^{-1}$. This peak may, however, be broadened and moved to a lower wavenumber, $3300\,\text{cm}^{-1}$, as in the case of ethanol, by hydrogen bonding.

Wavenumbers of the peaks characteristic of the —OH group are shown in figure 11.1 and figure 11.2 gives the infra-red absorption spectrum of ethanol.

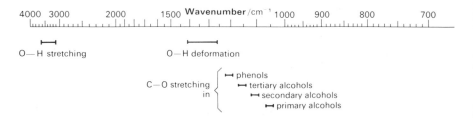

Figure 11.1
Typical wavenumbers for infra-red absorption by alcohols.

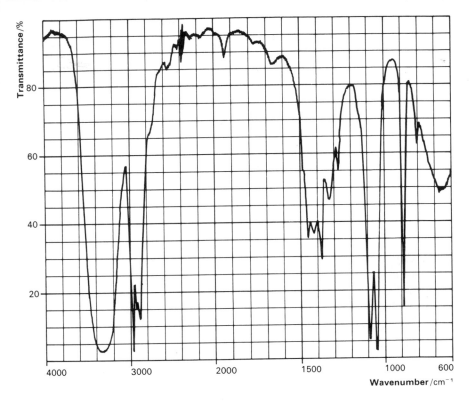

Figure 11.2
The infra-red spectrum of ethanol.

Water, like alcohols, contains the —OH group of atoms in its molecule, and so we may ask the question: how does the presence of an alkyl group instead of a hydrogen atom modify the properties of the hydroxyl group?

$$CH_3\text{—OH} \qquad H\text{—OH}$$

First consider bond energies:

$$E(CH_3\text{—OH}) = 336 \, kJ \, mol^{-1}$$

$$E(H\text{—OH}) \quad = 464 \, kJ \, mol^{-1}$$

If you look again at some of the bond energies quoted in Topic 9 you will appreciate that the H—OH bond is a strong bond but the CH₃—OH bond is somewhat less strong, comparable with the C—Hal bond in halogenoalkanes.

Dipole moments will give us an indication of possible bond polarity and again the values should be compared with those of halogenoalkanes.

Dipole moment/D

Water	1.85
Methanol	1.70
Ethanol	1.69
Octan-1-ol	1.68

The relative electronegativities of carbon and oxygen will result in a bond polarity $C^{\delta+}$—$O^{\delta-}$

Questions

1 Two important properties of water are its ability to ionize, and its ability to accept a proton, due to the presence of a lone pair of electrons on the oxygen atom.

$$H_2O(l) \rightleftharpoons H^+(aq) + OH^-(aq)$$

$$H_2O(l) + H^+(aq) \rightleftharpoons H_3O^+(aq)$$

Does the oxygen atom in alcohols also have a lone pair of electrons so that alcohols can accept protons as water does?

2 Consider the bonds in an alcohol:

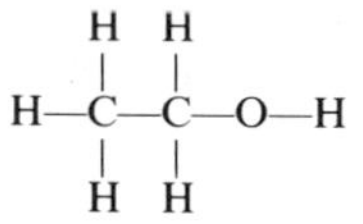

```
     H  H
     |  |
H—C—C—O—H
     |  |
     H  H
```

Which bonds are more likely to be broken in chemical reactions, C—H, C—O, or O—H?

3 Would you expect alcohols to take part most commonly in reactions with free radicals, electrophilic reagents, or nucleophilic reagents?

4 And what types of reaction might occur with alcohols: addition reactions, substitution reactions, elimination reactions?

To revise the terms in questions 3 and 4 look again at section 9.6 in Topic 9 if necessary.

Another interesting feature of alcohols is their high boiling points compared with those of their parent alkanes and corresponding halogenoalkanes.

$T_b/°C$		$T_b/°C$		$T_b/°C$	
Methane	−164	Chloromethane	−24	Methanol	+65
Ethane	−89	Chloroethane	+12	Ethanol	+78
Propane	−42	1-chloropropane	+47	Propan-1-ol	+97

What types of bonding might be responsible for this pattern of boiling points? Consider the types of bonding discussed in Topic 10.

Experiments with alcohols

The first three experiments are designed to be carried out on a range of alcohols. The remaining experiments can be tried out with any alcohol but are best carried out with ethanol because, as the carbon chain gets longer, the reactions get slower. You will be expected to carry out only a selection of the second group of experiments.

You should note that many of the substances used in these experiments are flammable and you should wear safety glasses throughout.

EXPERIMENT 11.1
An investigation of the reactions of the alcohols

Procedure for a range of alcohols

1 *Solubility in water* To 1 cm³ of ethanol in a test-tube add 1 cm³ of water. Do the two liquids mix? Test the mixture with Full-range Indicator.

Repeat with a range of alcohols, increasing the volume of water used if mixing does not occur.

What type of interaction would you predict between alcohol molecules and water molecules?

2 *Reaction with sodium* To 1 cm³ of ethanol in an evaporating basin add one small cube of freshly cut sodium. Is there any sign of reaction? Is reaction faster or slower or much slower than with water?

Repeat the experiment with a range of alcohols. Add ethanol to dissolve all traces of sodium before throwing away the reaction mixture.

Does this reaction appear comparable with the reaction between sodium and water? Which bond in the alcohol has been broken? What type of reaction has occurred?

3 *Oxidation* To a few cm³ of 2M sulphuric acid add a few drops of sodium dichromate(VI) solution. Next add 2 drops of ethanol and heat the reaction mixture until it just boils. Is there any sign of reaction? Is there any change of smell suggestive of a new organic compound? Repeat the experiment with a range of alcohols.

Procedure using ethanol alone

4 *A halogenoalkane from ethanol* Place about 5 cm³ of ethanol in a small pear-shaped flask and *cautiously*, in amounts of 0.5 cm³ at a time, add about 5 cm³ of concentrated sulphuric acid. Ensure that the contents of the flask are well mixed, and cool them by holding the flask under a running cold water tap. Quickly add 6 g of powdered potassium bromide and arrange the apparatus as shown in figure 11.3.

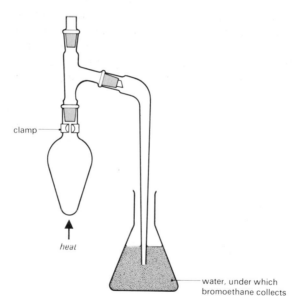

clamp

heat

Figure 11.3

water, under which bromoethane collects

Heat the flask by means of a low Bunsen burner flame, to distil the halogenoalkane which is made, and collect the product under water to ensure complete condensation. *Do not remove the flame without first removing the delivery tube*, or water may be sucked up into the hot flask. Next, take the conical flask, tip off the bulk of the water, and remove as much as possible of what remains, using a dropping pipette. The halogenoalkane may contain some hydrogen bromide which has been carried over during the distillation. Add about 20 cm³ of water to dissolve this, swirl the contents of the flask, and remove the water as before. The residual halogenoalkane will then be seen to be a colourless liquid, somewhat milky in appearance owing to the presence of water.

Test your product by putting a few drops into some 2M sodium hydroxide solution. Shake well and warm very gently. Acidify the solution with 2M nitric acid and add a few drops of 0.02M silver nitrate solution.

Compare your result with the result that you obtained in part 2 of experiment 9.3.

> Your product is bromoethane. Write an equation for the reaction and work out what bond was broken in the ethanol. What type of reaction occurred and what type of reagent attacked the ethanol?

5 *The dehydration of ethanol* Put ethanol in a test-tube to a depth of 1–2 cm. Push in some loosely packed ceramic fibre until all the ethanol has been soaked up. Now add a 2–3 cm depth of aluminium oxide granules and arrange the apparatus for the collection of a gas, as shown in figure 11.4.

Heat the granules gently and collect three or four test-tubes of gas, discarding the first one. Carry out the following tests on the gas:

a Test for combustion; any flame will be more visible if the test-tube is held upside down.

b Add 1–2 cm^3 of a mixture of equal volumes of 0.01M potassium manganate(VII) solution and 2M sulphuric acid. Shake the test-tube and look for any colour changes that suggest a reaction is occurring.

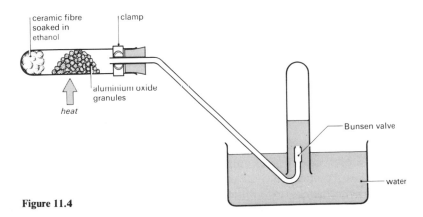

Figure 11.4

> What do the tests **a** and **b** tell you about the nature of the product from the dehydration of ethanol? What type of reaction is the dehydration of ethanol?

6 *An ester from ethanol* Put $2\,cm^3$ of ethanol in a test-tube and add $1\,cm^3$ of concentrated ethanoic acid and two or three drops of concentrated sulphuric acid (*TAKE CARE*). Warm the mixture gently in a hot water bath for five minutes, when an ester will be formed.

What does the product smell like? How does this compare with the smells of the starting materials? Pour the contents of the test-tube into a small beaker of sodium carbonate solution, to neutralize any excess of acid. Stir well and smell again. Is it like the smell of ethyl ethanoate from the bottle?

7 *The oxidation of ethanol to ethanoic acid* This experiment uses approximately equal quantities (0.02 mole) of ethanol and an oxidizing agent, and they are refluxed together in order to oxidize the ethanol as fully as possible under these conditions.

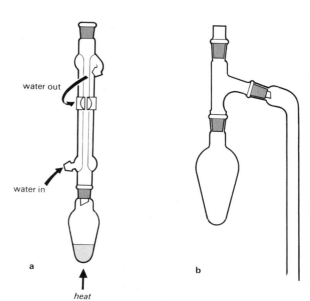

water out

water in

a

b

Figure 11.5

heat

Put about $10\,cm^3$ of 2M sulphuric acid in a 50-cm^3 pear-shaped flask and add about 6 g of sodium dichromate(VI) and 2–3 anti-bumping granules. Shake and warm the contents of the flask until solution is complete, and then *cool thoroughly* before adding, *with great care*, $2\,cm^3$ of concentrated sulphuric acid, to increase the concentration of the acid. *Cool again* and add about $1\,cm^3$ of ethanol *in drops* from a dropping pipette, pointing the flask away from your face. Put a condenser in the flask for reflux, as shown in figure 11.5a. Boil the contents of the flask gently, not allowing any vapour to escape, for 10 minutes. At the end of that time remove the Bunsen burner and arrange the apparatus for distillation, as shown in figure 11.5b. Gently distil 2–3 cm^3 of liquid into a test-tube.

The liquid that collects is an aqueous solution of ethanoic acid. Carry out the following tests on it, recording your results in tabular form. Leave two columns in your table in which to record the results of each test when performed on ethanol and ethanal.

a Notice the smell of the product.

b Will it neutralize an appreciable volume of sodium carbonate solution?

c Add a few drops of the product to 2 cm^3 of Fehling's solution (use equal volumes of solutions A and B) and boil the mixture. Fehling's solution contains a copper(II) compound and is used to test for organic reducing agents.

Compare these results with those obtained using ethanol, and ethanal (made in the next experiment).

8 *The oxidation of ethanol to ethanal* This experiment uses only half the quantity of oxidizing agent (0.01 mole) that the previous experiment used and the product is distilled from the reaction mixture immediately it is formed. In this way we hope to achieve a partial oxidation of ethanol.

Place about 10 cm^3 of 2M sulphuric acid in a flask and add about 3 g of sodium dichromate(VI) and 2–3 anti-bumping granules. Shake the contents of the flask until solution is complete (warming if necessary but cooling afterwards). Cool the mixture and add about 1 cm^3 of ethanol in drops from a dropping pipette, shaking the flask so as to mix the contents, and then assemble the apparatus as in figure 11.5b. Gently distil 2–3 cm^3 of liquid into a test-tube, taking care that none of the reaction mixture splashes over. The product is an aqueous solution of ethanal.

Carry out the tests 7a, b, and c, above, comparing the results.

Reactions of alcohols

1 *Hydrogen bonding* In alcohols such as ethanol a very high proportion of the molecules are hydrogen-bonded. Hydrogen bonding between ethanol molecules and water molecules accounts for the unusual behaviour of ethanol–water

mixtures when distilled. All ethanol–water mixtures when fractionally distilled produce a mixture whose composition cannot be altered by boiling. This 'constant-boiling mixture' contains 4% water and 96% ethanol. Anhydrous ethanol, or 'absolute' alcohol, can only be obtained by the use of an efficient drying agent like calcium oxide or, in industry, the addition of benzene to disrupt the hydrogen-bonded system, followed by distillation.

2 *Acid-base properties* The reaction with sodium metal suggests that ethanol can act as a weak acid in its reaction with sodium.

$$2Na + 2CH_3CH_2OH \longrightarrow 2CH_3CH_2O^-Na^+ + H_2$$

ethanol   sodium ethoxide

It is a weaker acid than water, and in the liquid state has no appreciable ionization, and no significant reaction with sodium hydroxide.

Water can also act as a base, accepting protons, and so can ethanol.

$$H^+ + CH_3CH_2OH \rightleftharpoons CH_3CH_2OH_2^+$$

Write the two parallel equations for water and note that the pattern of behaviour is comparable.

3 *Oxidation to form a C$=$O group* When primary alcohols are treated with oxidizing agents such as acidified sodium dichromate(VI), aldehydes may be obtained, provided they are not exposed to further oxidation.

$$3CH_3CH_2OH + Cr_2O_7^{2-} + 8H^+ \longrightarrow 3CH_3CHO + 2Cr^{3+} + 7H_2O$$

ethanol   ethanal

Under more vigorous conditions and with an additional quantity of oxidizing agent, carboxylic acids are obtained.

$$3CH_3CH_2OH + 2Cr_2O_7^{2-} + 16H^+ \longrightarrow 3CH_3CO_2H + 4Cr^{3+} + 11H_2O$$

ethanol   ethanoic acid

If secondary alcohols are treated with oxidizing agents ketones are obtained. The reaction usually proceeds no further because the formation of a carboxylic acid would require a strong C—C bond to be broken.

$$3CH_3CHOHCH_3 + Cr_2O_7^{2-} + 8H^+ \longrightarrow 3CH_3COCH_3 + 2Cr^{3+} + 7H_2O$$

propan-2-ol   propanone

Tertiary alcohols resist oxidation because the reaction would require a strong C—C bond to be broken even to form an aldehyde or a ketone.

Oxidation number change can be used to work out the relative amounts of sodium dichromate(VI) needed in these oxidations. Follow the usual procedure of taking the oxidation number of oxygen as -2 and that of hydrogen as $+1$. What is the oxidation number of carbon in each compound of the following sequence?

$$
\underset{\text{methanol}}{\overset{\displaystyle H}{\underset{\displaystyle H}{\overset{\displaystyle |}{\underset{\displaystyle |}{H-C-O-H}}}}} \longrightarrow \underset{\text{methanal}}{\overset{\displaystyle H}{\underset{\displaystyle H}{{}^{\diagdown}C{=}O}}} \longrightarrow \underset{\text{methanoic acid}}{\overset{\displaystyle H}{\underset{\displaystyle H-O}{{}^{\diagdown}C{=}O}}}
$$

You should find that the oxidation number of carbon changes by $+2$ at each stage. This has to be balanced by a change of -3 when dichromate(VI) ions change to chromium(III) ions.

What is the correct ratio of moles of methanol and dichromate(VI) to mix for oxidation through one stage, and through two stages?

4 Nucleophilic substitution at the carbon atom to form halogenoalkanes
Different procedures are adopted for the different halogenoalkanes, depending on their relative reactivity and the halogen compounds available, but the overall reaction can be expressed as:

$$ROH + HHal \rightleftharpoons RHal + H_2O$$

Conditions are usually selected to be strongly acidic. Under these conditions the alcohol is first protonated to form an intermediate (not isolated) which is attacked by the nucleophile Br^-. Protonation results in the *leaving group* being water rather than the hydroxide ion.

$$CH_3CH_2-O-H \xrightarrow[\text{protonation}]{H^+} CH_3CH_2-\overset{+}{\underset{\displaystyle H}{\overset{\displaystyle H}{O}}} \xrightarrow[\text{nucleophile}]{Br^-} CH_3CH_2-Br + H_2O$$

Depending on the structure of the alcohol and the choice of halogen, more or less severe conditions are necessary: in part 4 of experiment 11.1 concentrated sulphuric acid reacted with potassium bromide to produce hydrogen bromide, but it was also needed to make the conditions sufficiently acidic.

Substitution of alcohols by chlorine can also be achieved by the use of phosphorus pentachloride.

5 *Dehydration to alkenes* When alcohol vapours are passed over a catalyst such as aluminium oxide at about $400\,°C$, they are dehydrated to alkenes. Alternatively, concentrated phosphoric or sulphuric acid can be used as a dehydrating agent.

$$CH_3CH_2OH \longrightarrow CH_2{=}CH_2 + H_2O$$
ethanol   ethene

This elimination reaction is the reverse of the hydration of alkenes that was described in Topic 9, section 9.4.

6   *Formation of esters* The reaction of an alcohol with a carboxylic acid forms an ester and water.

$$RCO_2H + R'OH \rightleftharpoons RCO_2R' + H_2O$$

Reaction is very slow unless a small amount of a strong acid is added as a catalyst.
   We shall consider this reaction in more detail in Topic 13. The reaction is a nucleophilic substitution at the carbon atom of the functional group of the acid, with the alcohol acting as the nucleophile.

11.2
PHENOL

In alcohols the hydroxyl group, —OH, is attached to an alkyl group. You are now going to examine phenol, a compound in which the molecule has a hydroxyl group attached to a benzene ring.

$$CH_3{-}CH_2{-}OH$$

—OH

ethanol   phenol

   In these experiments your principal objective will be to find out if the properties of the hydroxyl group are modified by being attached to a benzene ring instead of an alkyl group. You will thus see some of the characteristic features of the chemistry of phenol.

EXPERIMENT 11.2
An investigation of the reactions of phenol

   WARNING: Phenol is corrosive. Both crystals of phenol and solutions containing it will cause unpleasant blisters if allowed to come in contact with your skin. *You must handle it very carefully. You must also wear safety glasses.*

Procedure

1 *Solubility in water* Put about 5 cm³ of water in a test-tube and add a few crystals of phenol. Notice the characteristic smell of phenol.

Are there any oily drops left in the water?
Does phenol dissolve in water?

Test the phenol–water mixture with Full-range Indicator.

What is the pH of the solution?

Compare the results with the effect of an ethanol–water mixture on Full-range Indicator.

2 *Phenol as an acid*

a *Action of sodium* Place some crystals of phenol in a dry test-tube and warm them until they are molten. Add a small cube (2–3 mm side) of sodium and watch carefully. *Do not heat the tube continuously.* What do you observe? Dispose of the contents with care: do *not* pour down the sink in case some sodium remains but carefully add 1–2 cm³ of ethanol to the *cool* test-tube, wait until all bubbling has ceased, and then pour away.

What bond has been broken in phenol, a C—H bond or an O—H bond?
Compare this reaction with that of ethanol and sodium.

b *Action of sodium hydroxide* To about 5 cm³ of 2M sodium hydroxide solution in a test-tube add a few crystals of phenol, and compare the solubility in alkali with the solubility in water. Now add about 2 cm³ of concentrated hydrochloric acid.

What do you observe and what does this tell you?

c *Action of sodium carbonate* To about 5 cm³ of M sodium carbonate solution add a few crystals of phenol.

Is there an effervescence of carbon dioxide gas? Does this suggest that phenol is a strong or a weak acid?

3 *Formation of an ester* Place about 0.5 g of phenol in a test-tube and add 4 cm³ of 2M sodium hydroxide to dissolve the phenol. Add 1 cm³ of ethanoic anhydride, cork the test-tube, and shake it for a few minutes.

An emulsion of phenyl ethanoate should form. Note the characteristic smell of the ester.

In what ways did the conditions for this reaction differ from the preparation of ethyl ethanoate in experiment 11.1 part 6?

4 *Properties of the benzene ring*

a *Combustion* Set fire to a small crystal of phenol on a combustion spoon.

What sort of flame do you see? Is a similar flame obtained when ethanol is burned?

Compare these results with your results for the combustion of alkanes and arenes in Topic 9. Does the presence of a hydroxyl group alter the result?

b *Iron(III) chloride* Dissolve two or three small crystals of iron(III) chloride in about $5\,cm^3$ of water, and add about $1\,cm^3$ of a solution of phenol in water. Note the intense colour that is formed.

The formation of intense colours is characteristic of several compounds having hydroxyl groups attached to benzene rings. Find out if this also happens with ethanol.

c *Bromine* To a solution of phenol in water add bromine water.

How readily does reaction occur?

d *Nitric acid* To a solution of phenol in water add 2M nitric acid.

How readily do coloured products appear?

Compare your results with those obtained using methoxybenzene in experiment 9.5a.

Do you think these are substitution reactions? Has phenol been attacked by electrophilic reagents?

A comparison of phenol with ethanol

1 *Phenol as an acid* Phenol loses a proton much more readily than ethanol; they both form a sodium salt by reaction with sodium metal, but only phenol will neutralize sodium hydroxide to form its sodium salt.

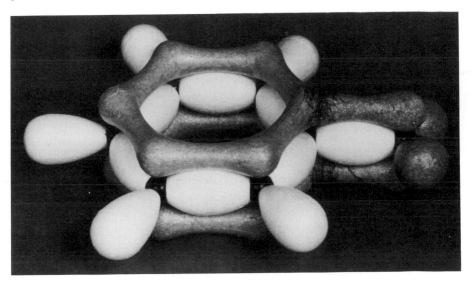

$$\text{C}_6\text{H}_5\text{—OH} + \text{Na}^+\text{OH}^- \longrightarrow \text{C}_6\text{H}_5\text{—O}^-\text{Na}^+ + \text{H}_2\text{O}$$

sodium phenoxide

It is suggested that the phenoxide ion is formed much more readily than the ethoxide ion because one of the lone pairs of electrons on the oxygen atom in phenol can join the delocalized electrons of the benzene ring. In this way, the negative charge of the ion is stabilized by being spread out over the whole molecule. The ethoxide ion cannot form a delocalized system and so cannot gain the same stability as the phenoxide ion.

Figure 11.6
A PEEL model showing the delocalized orbitals in the phenoxide ion.
Model, Griffin & George Ltd; photograph, University of Bristol, Faculty of Arts Photographic Unit.

The delocalization of electrons in the benzene ring was described in Topic 9, page 303.

Phenol does not, however, react with the weak base, sodium carbonate, to produce an effervescence of carbon dioxide.

So phenol is only a weak acid; the pH of its solution in water is only slightly less than 7.

2 *Phenol as an alcohol*
a The interaction of a lone pair of electrons on the oxygen atom of phenol

with the benzene ring means that phenol normally forms esters much less readily than ethanol.

carboxylic acid + alcohol $\longrightarrow$ ester + water

Instead of using a carboxylic acid with an acid catalyst as in the case of ethanol, the more reactive carboxylic acid anhydrides or the acyl chlorides (see Topic 13) must be used.

However, in the esterification reaction the alcohol is reacting as a nucleophile, $CH_3CH_2\overset{\cdot\cdot}{O}H$. In the case of phenol the reactivity can be enhanced by the addition of sodium hydroxide to form the phenoxide ion, ⎔—$\overset{\cdot\cdot}{O}^-$, which is a more reactive nucleophile than phenol itself, ⎔—$\overset{\cdot\cdot}{O}H$.

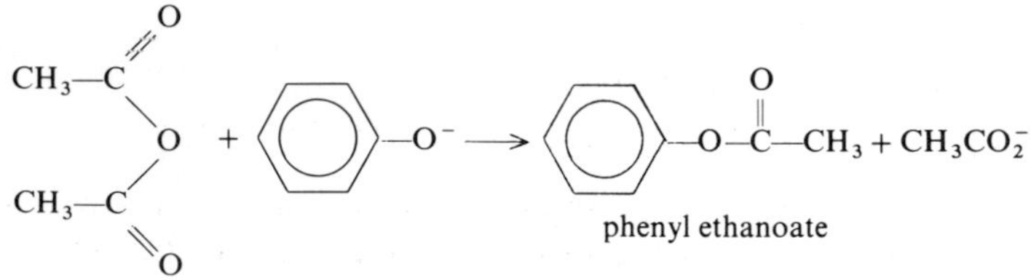

phenyl ethanoate

b Phenol will not undergo nucleophilic substitution of the hydroxyl group to form halogenoarenes, whereas ethanol readily reacts with the nucleophile HBr to form bromoethane.

$$CH_3CH_2OH + HBr \longrightarrow CH_3CH_2Br + H_2O$$

In phenol the interaction of the lone pair of electrons on the oxygen atom with the benzene ring makes it difficult to break the C—O bond.

c Oxidation of the hydroxyl group in phenol to a carbonyl group cannot be carried out because that would involve disruption of the stable benzene ring.

3 *Phenol as an arene* The delocalization of electrons from the oxygen in phenol into the benzene ring makes phenol more susceptible than benzene to substitution reactions with electrophilic reagents such as bromine and nitric acid. The reactions occur so readily that multiple substitution often occurs.

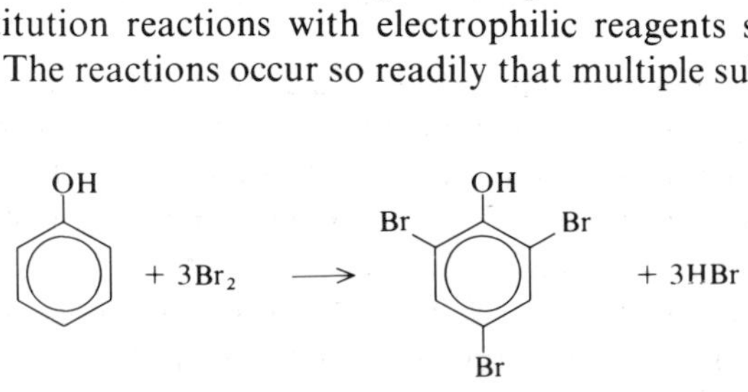

2,4,6-tribromophenol

2-nitro-
phenol   4-nitrophenol

2,4,6-trinitrophenol
(picric acid)

BACKGROUND READING 1
The importance of phenol and its derivatives

Phenol has given its name to all compounds which have a hydroxyl group attached to a benzene ring; they are known collectively as 'the phenols'.
Examples include the three isomers:

2-methylphenol   3-methylphenol   4-methylphenol

These compounds are known as *cresols*, and the three isomers have alternative names which you may see in other books: *ortho*-cresol, *meta*-cresol, and *para*-cresol. The prefixes, *ortho*, *meta*, and *para* are an older terminology for compounds having two substituents attached to the same benzene ring; *ortho* (*o*-), if the substituents are attached to adjacent carbon atoms, *meta* (*m*-), if they are separated by one carbon atom, and *para* (*p*-), if they are separated by two carbon atoms.

Phenol itself is an industrial raw material of considerable importance.

Its use as an antiseptic, first adopted by Lister in the 1860s, is widely known and it has been publicized as a component of soap since the last century, for instance in 'Pears Carbolic Soap' and 'Wright's Coal Tar Soap'. Far more important currently is its use in the production of phenolic resins used in the manufacture of plastics such as Bakelite. This, and the other major uses of phenol, are listed at the end of this section; the laboratory preparation of a phenolic resin is described in Topic 17.

A number of substituted phenols have important uses. One interesting example is their use as germicides in proprietary products such as Dettol and TCP, a modern development of Lister's discovery:

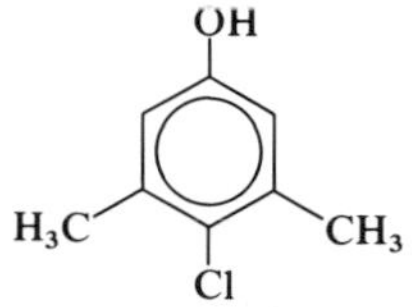

4-chloro-3,5-dimethylphenol (Dettol) 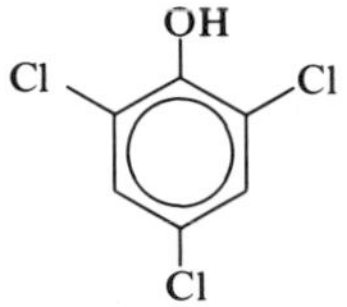   2,4,6-trichlorophenol (TCP)

The effectiveness of germicides can be expressed as a phenol coefficient:

$$= \frac{\text{concentration of phenol required to kill the germs}}{\text{concentration of germicide required to kill the same germs}}$$

Some values obtained using the 'germ' *Salmonella typhosa* are given below.

Germicide	Phenol coefficient
Phenol	1
2-chlorophenol	4
2,4-dichlorophenol	13
TCP	23
Dettol	280

The history of the manufacture of phenol provides an interesting example of how social and financial considerations influence the choice of chemical reactions used in industrial processes.

The original large-scale source of phenol was coal tar, which contains some two hundred organic compounds. Phenol occurs in the 'middle oil' fraction of coal tar. This is obtained by fractional distillation of coal tar, collecting the fraction which boils over the range 170–230 °C. Phenol can be isolated from this by extraction with alkali and fractional distillation.

The First World War brought such a sudden large increase in the demand for phenol that the supplies from coal tar were no longer enough. Phenol was required for conversion to 2,4,6-trinitrophenol (picric acid) which was used as an explosive, and to cope with this sudden increased demand the first synthetic method for phenol manufacture was introduced.

2,4,6-trinitrophenol

In this method, benzene was sulphonated, and the benzenesulphonic acid thus produced was converted to phenol by fusing with alkali.

Immediately after the First World War, large stocks of phenol had been synthesized and the demand suddenly fell off, with the result that production stopped and prices dropped. However, a new demand for phenol was shortly to come from a new industry, the manufacture of plastics, using phenol–methanal polymers. The stock remaining from war production was rapidly consumed, but, as this had been sold at less than cost price, a cheaper and more efficient method of manufacture was needed to prevent a severe rise in prices.

This resulted in the discovery of a new process by the Dow Chemical Company in America. In this, chlorobenzene is heated with aqueous sodium hydroxide under pressure at 300 °C.

The Dow process had the disadvantage of using up chlorine, which has to be manufactured, and yielding as its only by-product sodium chloride, a common naturally occurring substance.

These disadvantages were removed in the Raschig Process, developed in Germany. This is a two-stage process. In the first stage benzene, hydrogen chloride, and air are passed over a catalyst of copper(ii) chloride at 200 °C, to produce chlorobenzene and water. In the second stage the chlorobenzene and steam are passed over a second catalyst of silicic acid at 400 °C, and react to give phenol and hydrogen chloride.

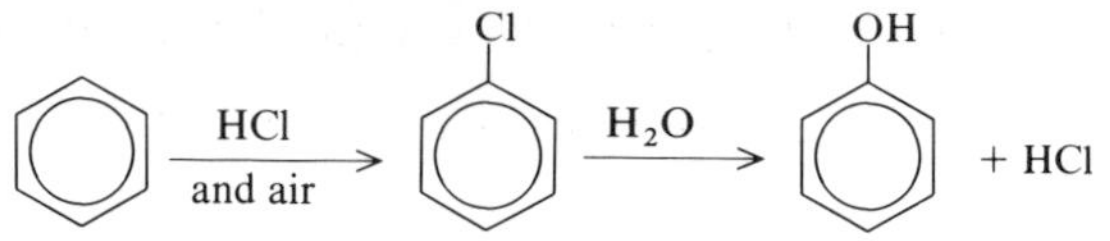

It can be seen that in the overall reaction the only substances actually used up are benzene and air, because the hydrogen chloride used in the first stage is regenerated in the second.

During and after the Second World War petroleum became a very important source of raw materials for the chemical industry, and the petrochemical industry has grown to be an industry in its own right. A process for manufacturing phenol from petroleum products was developed by the Distillers Company in Great Britain.

Propene, from the steam cracking of naphtha, from crude oil, is made to react with hot benzene in the presence of an aluminium chloride catalyst, forming (1-methylethyl)benzene (cumene). This, on reaction with oxygen at 110 °C, gives a peroxide which is decomposed by acid to give phenol and propanone. The products can be sold in equal tonnages, making the process highly competitive.

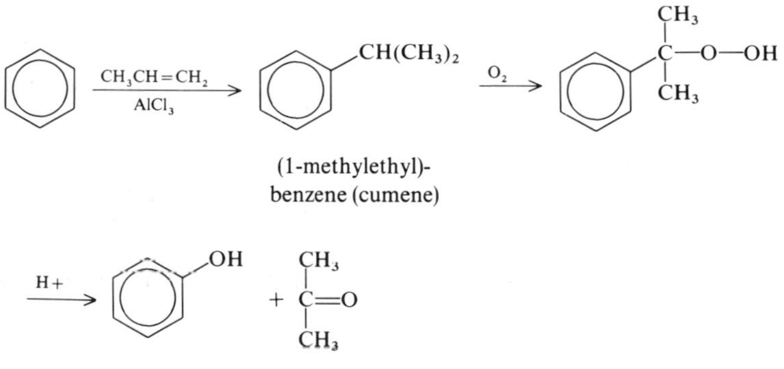

The process is generally referred to as the cumene oxidation process. It has almost completely replaced all the other synthetic processes for the manufacture of phenol.

The major industrial uses of phenol at the present time include the manufacture of the following:

Phenolic resins including phenol–methanal (phenol–formaldehyde or PF resins)
Diphenylolpropane (bisphenol A) for epoxy resins and polycarbonates
Caprolactam for nylon production
Phenolic plasticizers for use in poly(chloroethene) (poly(vinyl chloride)) materials
Alkylphenols for non-ionic detergents.

It can be seen that the majority of the phenol that is produced ends up in synthetic polymers or fibres.

11.3
CARBONYL COMPOUNDS

Carbonyl compounds have the unsaturated functional group $\diagdown C = O$. They are subdivided into two closely related types of compound, the aldehydes and the ketones.

Compounds with one alkyl group and one hydrogen atom attached to the carbonyl group are known as *aldehydes*. They are named after their parent alkane, with the terminal *e* replaced by *al*.

Compounds with two alkyl groups attached to the carbonyl group are known as *ketones*. They are named in the same manner, but with the terminal *e* replaced by *one*.

Examples of these two types of compound are:

	Aldehyde	Ketone
a Displayed formulae	H O | // H—C—C | \ H H	H O H | || | H—C—C—C—H | | H H
b Structural formulae	CH_3—CHO ethanal	CH_3—CO—CH_3 propanone

Space-filling models of the molecules of these two compounds are shown in figure 11.7.

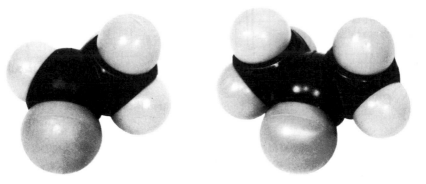

Figure 11.7
Space-filling models of the molecules of ethanal and propanone.

propanal, CH_3CH_2—CHO, boils at 48 °C; and propanone, CH_3—CO—CH_3, boils at 56 °C.

For our experiments we shall need only one aldehyde and one ketone, and we shall use propanal and propanone, which are relatively easy to handle, although highly flammable. You should exercise care, including the wearing of safety glasses.

Propanone used to be known as acetone. This name is still widely used.

EXPERIMENT 11.3
An investigation of the reactions of aldehydes and ketones

Procedure

1 *Solubility in water* To 1 cm^3 of propanone in a test-tube add 1 cm^3 of water. Do the two liquids mix? Repeat with propanal.

What type of interaction would you predict between carbonyl groups and water molecules?

2 *Addition reaction with hydrogen sulphite ions* To a few drops of a carbonyl compound add 5 cm^3 of a fresh saturated solution of sodium disulphate(IV) and shake well. What signs of reaction are there?

Sodium disulphate(IV), $Na_2S_2O_5$, dissolves in water, forming hydrogen sulphite ions, HSO_3^-.

Are hydrogen sulphite ions likely to react as electrophiles or nucleophiles?

Will hydrogen sulphite ions attack the carbon atom or the oxygen atom of the carbonyl group?

3 *Addition–elimination reaction with Brady's reagent* To a few drops of a carbonyl compound in a test-tube add 5 cm^3 of Brady's reagent. Note the formation of a coloured crystalline solid.

Brady's reagent contains a nucleophile, 2,4-dinitrophenylhydrazine, dissolved in methanol–water and concentrated sulphuric acid.

The importance of this reagent is discussed in Topic 17.

4 *Reaction with halogens* To $5\,cm^3$ of $0.01M$ iodine add $1\,cm^3$ of propanone, divide into two portions, and to one portion add an equal volume of $2M$ sulphuric acid. Stand the reaction mixtures in a hot water bath and see if the iodine colour fades. Which is likely to have reacted, the carbonyl group or the alkyl groups?

5 *Oxidation*
a To a few cm^3 of $2M$ sulphuric acid in a test-tube add a few drops of potassium (or sodium) dichromate(VI) solution. Next add 2 drops of a carbonyl compound (*CARE:* flammable) and heat the mixture until it *just* boils. Try to tell from any colour changes of the dichromate(VI) ion if it oxidizes the carbonyl compound.

Is there any difference in the behaviour of aldehydes and ketones?

b Mix equal volumes of Fehling's solutions A and B, add a few drops of a carbonyl compound, and heat in a boiling water bath for three minutes. Note any colour changes and the formation of any precipitate.

Which carbonyl compounds react with Fehling's solution, aldehydes or ketones?

6 *Laboratory preparation: reduction of 1,2-diphenylethanedione* Although this preparation involves the use of rather unusual chemicals the reaction is essentially the reduction of carbonyl groups to hydroxyl groups by hydride ions (H^-).

Sodium tetrahydridoborate(III), $NaBH_4$, is used as a convenient source of hydride ions. A convenient ketone to use is 1,2-diphenylethanedione, because both it and the product of the reaction, 1,2-diphenylethanediol, are solids and therefore easily handled.

Add $0.7\,g$ of 1,2-diphenylethanedione to $7\,cm^3$ of 95% aqueous ethanol in a conical flask and warm to dissolve the solid. Allow to cool, when the solid will reappear as a fine suspension. Now add $0.15\,g$ of sodium tetrahydridoborate(III) and allow to stand for ten minutes. Heat should be evolved. Add $15\,cm^3$ of water and heat to boiling to destroy any excess of sodium tetrahydridoborate(III). On cooling, crystals of 1,2-diphenylethane-1,2-diol (T_m $139\,°C$) should be formed. Collect the crystals by suction filtration, continuing to draw air through the filter until the crystals are nearly dry. Transfer them to a clean piece of filter paper and allow them to dry in the air. Weigh your product and calculate the yield as a percentage of the expected mass of product.

Draw the structural formula of the product and carry out appropriate experiments on the starting material and your product to test for their functional groups. If you have time determine their melting points, using the procedure described in Topic 9, page 310.

Reactions of aldehydes and ketones

1 *With water* Carbonyl compounds with small alkyl groups mix readily with water owing to hydrogen bonding. The oxygen atom has two unshared pairs of electrons available and the possibility of hydrogen bonding is enhanced by the polarity of the bond.

$$
\begin{array}{c}
\text{O—H}\cdots\text{O}=\text{C} \overset{\displaystyle \text{CH}_3}{\underset{\displaystyle \text{CH}_3}{}} \\
\text{H}
\end{array}
$$

The interaction of aldehydes with water can go much further than hydrogen bonding, and water will add to the double bond.

$$
\underset{\text{ethanal}}{\text{CH}_3-\overset{\displaystyle \text{H}}{\text{C}}=\text{O}} + \text{H}_2\text{O} \rightleftharpoons \underset{\text{'ethanal hydrate'}}{\text{CH}_3-\overset{\displaystyle \text{H}}{\underset{\displaystyle \text{OH}}{\text{C}}}-\text{OH}}
$$

Methanal is almost totally hydrated in aqueous solution, and ethanal to the extent of 58 %, but the presence of two methyl groups in propanone reduces its hydration to a very small proportion. The reaction is readily reversible and hydrates can only be isolated in exceptional cases.

2 *Addition reactions*
a Sodium disulphate(IV) dissolves in water to give mainly sodium hydrogen sulphite:

$$
\text{Na}_2\text{S}_2\text{O}_5(\text{s}) + \text{H}_2\text{O}(\text{l}) \longrightarrow 2\text{Na}^+(\text{aq}) + 2\text{HSO}_3^-(\text{aq})
$$

A saturated solution of sodium hydrogen sulphite will react with aldehydes (—CHO) and methyl ketones (CH_3—CO—) to form crystalline products.

$$CH_3-CH_2-CHO + Na^+HO-\overset{\overset{\displaystyle O}{\|}}{\underset{\underset{\displaystyle O^-}{|}}{S:}}$$

$$\longrightarrow CH_3-CH_2-CH(OH)-SO_3^- Na^+$$

The sulphur atom provides the electron pair for a nucleophilic addition to the aldehyde.

b When an aqueous mixture of a carbonyl compound and sodium cyanide is gradually acidified an addition reaction occurs. The product is a hydroxynitrile (cyanohydrin) which is a useful intermediate in synthesis because the number of carbon atoms in the molecule has been increased (see Topic 17).

$$CH_3-CO-CH_3 + NaCN \xrightarrow{H^+} H_3C-\overset{\overset{\displaystyle OH}{|}}{\underset{\underset{\displaystyle CN}{|}}{C}}-CH_3$$

2-hydroxy-2-methyl-
propanenitrile

The reaction only occurs in mildly acidic conditions.

The reaction begins with the nucleophilic addition of a cyanide ion, :CN⁻.

This occurs because of the formation of —O⁻ from $\overset{\diagdown}{\underset{\diagup}{C}}{}^{\delta+}=O^{\delta-}$. If pH is too

low the :CN⁻ ions form molecules of the weak acid HCN, and the nucleophilic addition does not occur.

$$CH_3-\overset{\overset{\displaystyle O}{\|}}{\underset{\underset{\displaystyle CH_3}{|}}{C}} + :CN^- \longrightarrow \left[CH_3-\overset{\overset{\displaystyle O^-}{|}}{\underset{\underset{\displaystyle CH_3}{|}}{C}}-CN \right]$$

The reaction continues with the protonation of the intermediate, either by H⁺ or by solvent, H₂O.

$$\left[CH_3-\overset{\overset{\displaystyle O^-}{|}}{\underset{\underset{\displaystyle CH_3}{|}}{C}}-CN \right] + H_2O \longrightarrow CH_3-\overset{\overset{\displaystyle OH}{|}}{\underset{\underset{\displaystyle CH_3}{|}}{C}}-CN + OH^-$$

3 *Addition–elimination at the double bond with Brady's reagent*   The reaction of Brady's reagent, 2,4-dinitrophenylhydrazine, with carbonyl compounds has a mechanism which involves an initial nucleophilic addition, immediately followed by an elimination reaction.

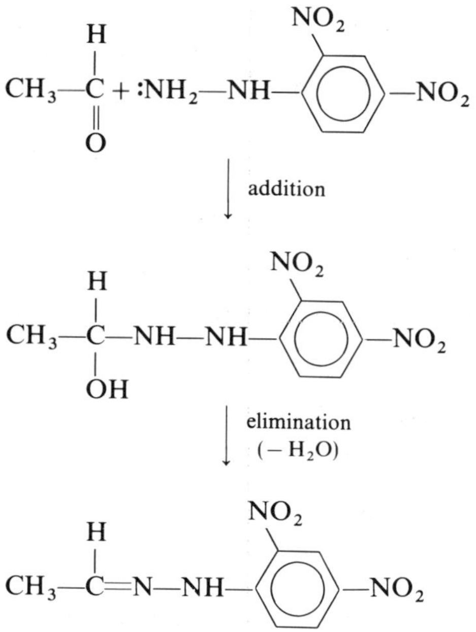

$$CH_3-\overset{\overset{\displaystyle H}{|}}{\underset{\underset{\displaystyle O}{\|}}{C}} + :NH_2-NH-\underset{}{\bigcirc}-NO_2 \quad (NO_2)$$

addition

$$CH_3-\overset{\overset{\displaystyle H}{|}}{\underset{\underset{\displaystyle OH}{|}}{C}}-NH-NH-\underset{}{\bigcirc}-NO_2 \quad (NO_2)$$

elimination
$(-H_2O)$

$$CH_3-\overset{\overset{\displaystyle H}{|}}{C}=N-NH-\underset{}{\bigcirc}-NO_2 \quad (NO_2)$$

ethanal 2,4-dinitro-
phenylhydrazone

4 *Halogens* The reaction of ketones with halogen in the presence of acid or base catalyst results in substitution on the carbon atom adjacent to the carbonyl group. The reaction will be investigated in more detail in Topic 14.

$$CH_3-CO-CH_3 + I_2 \xrightarrow{H^+} CH_3-CO-CH_2I + HI$$

This substitution reaction should be contrasted with the addition reaction of alkenes with halogens, Topic 9, page 296. Aldehydes cannot be halogenated directly because the aldehyde group is oxidized by halogens.

5 *Oxidation*
a The dichromate(VI) ion and other mild oxidizing agents convert aldehydes to carboxylic acids; ketones cannot be oxidized without breaking a carbon–carbon bond.

$$3CH_3CHO + Cr_2O_7^{2-} + 8H^+ \longrightarrow 3CH_3CO_2H + 4H_2O + 2Cr^{3+}$$

ethanal   ethanoic
acid

b *Fehling's solution* When mixed, Fehling's solution contains an unstable copper(II) complex ion in alkaline solution. The complex is with the anion of 2,3-dihydroxybutanedioic acid. Most aldehydes reduce the Cu(II) ions to Cu(I) ions, resulting in the precipitation of the brick-red coloured copper(I) oxide. Ketones do not react.

One application of this test is described in the Background reading on 'Diabetes' (pages 406–8).

6 *Reduction* Hydrogen at ordinary temperature and high pressure will react with carbonyl compounds if a nickel catalyst is used, producing alcohols in excellent yield. Aldehydes give primary alcohols and ketones give secondary alcohols.

$$CH_3CH_2CHO + H_2 \longrightarrow CH_3CH_2CH_2OH$$

propanal   propan-1-ol

$$CH_3COCH_3 + H_2 \longrightarrow CH_3CHOHCH_3$$

propanone   propan-2-ol

Sodium tetrahydridoborate(III) is a useful laboratory reagent. It reduces carbonyl compounds by transferring the nucleophile H^- to the carbon atom of the carbonyl group, and the reaction is therefore analogous to the addition reactions described above.

$$CH_3-\overset{\overset{\displaystyle O}{\|}}{C}-H \xrightarrow[\text{(NaBH}_4)]{H^-} \left[CH_3-\overset{\overset{\displaystyle O^-}{|}}{\underset{\underset{\displaystyle H}{|}}{C}}-H \right] \xrightarrow[\text{acid}]{H^+} CH_3CH_2OH$$

11.4
CARBOHYDRATES

Carbohydrates are so named because they have the general formula $C_n(H_2O)_m$. As you might expect, this formula does not accurately represent their structures, which usually involve a number of alcoholic hydroxyl groups plus an aldehyde or ketone carbonyl group. We shall be considering the chemistry of four important carbohydrates, glucose, fructose, sucrose, and starch.

Glucose, $C_6H_{12}O_6$, is classified as an *aldohexose*: *aldo* denoting an aldehyde functional group, *hex* denoting six carbon atoms, and *ose* denoting that it is a carbohydrate. The chain form of glucose is shown below.

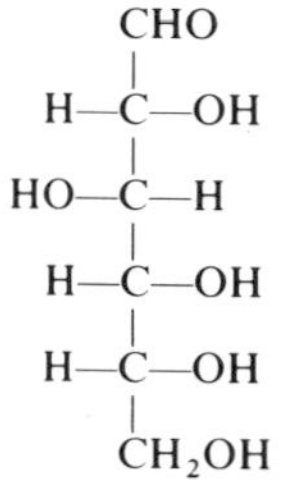

```
        CHO
         |
    H—C—OH
         |
  HO—C—H
         |
    H—C—OH
         |
    H—C—OH
         |            glucose
    CH₂OH      (chain form)
```

Other aldohexoses have different arrangements of their hydroxyl groups.

This, however, is not an entirely satisfactory way to represent the structure of glucose because the aldehyde group is normally linked to the C_5 carbon atom in a six-membered ring. The ring form exists as two isomers, as shown in the following *stereochemical formulae*:

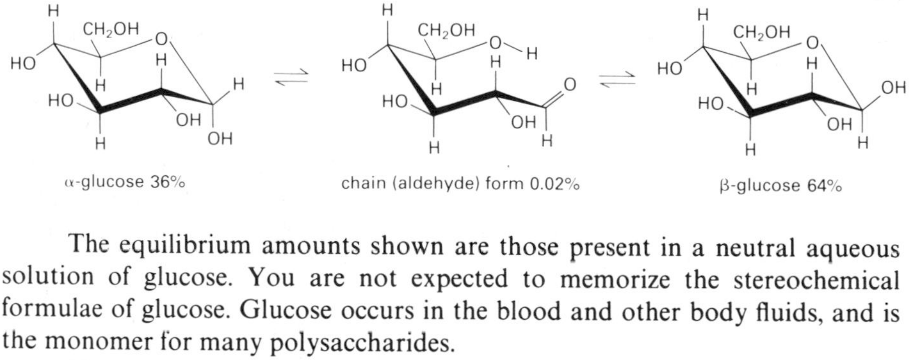

α-glucose 36%   chain (aldehyde) form 0.02%   β-glucose 64%

The equilibrium amounts shown are those present in a neutral aqueous solution of glucose. You are not expected to memorize the stereochemical formulae of glucose. Glucose occurs in the blood and other body fluids, and is the monomer for many polysaccharides.

Fructose, $C_6H_{12}O_6$, is a *ketohexose* because it has a ketone functional group.

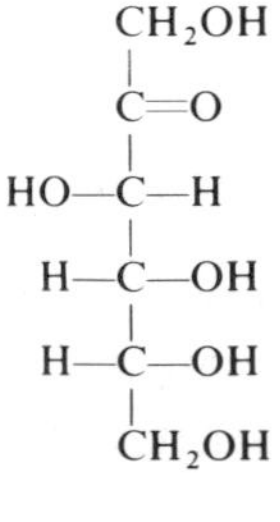

```
        CH₂OH
         |
         C=O
         |
  HO—C—H
         |
    H—C—OH
         |
    H—C—OH
         |            fructose
    CH₂OH      (chain form)
```

Fructose also forms a ring structure with a keto group linked to the C_5 carbon atom. Fructose forms a five-membered ring whereas glucose forms a six-membered ring.

fructose

Sucrose, $C_{12}H_{22}O_{11}$, is the common sugar of our diet and is obtained from sugar cane or sugar beet. It can be regarded as the combination of one unit of glucose plus one unit of fructose.

sucrose

Glucose and fructose are classified as monosaccharides, while sucrose is a disaccharide as it is made of two monosaccharide units.

Other carbohydrates such as cellulose and starch are polysaccharides, and are built up from several hundred monosaccharide units. They are used by plants and animals mainly as structural material and as food reserves. Thus *cellulose*, a polymer of as many as five thousand glucose units, forms the framework for cells in plant tissue. *Starch* is the main food reserve of plants, forming up to 80% of the mass of seeds, and consists of up to three thousand sub-units of glucose. Cellulose and starch differ in being polymers of the two different isomers of glucose; cellulose is the polymer of β-glucose and starch is the polymer of α-glucose. The shell of a crab and the outer covering of a beetle are made of *chitin*, another polysaccharide, but from a different monomer.

Chirality

Look at your hands, keeping the palms uppermost. You cannot cover one hand with the other, because one is a left hand and the other is a right hand. Now compare your left palm with the reflection of your right palm in a mirror. You will now be looking at what appear to be two left palms.

When two objects exist which are identical in appearance only through a mirror image they are said to be *chiral* (from the ancient Greek for 'hand', pronounced *kiral*). Like familiar objects such as gloves and shoes, some compounds also possess the property of being chiral.

B solutions. Bring to the boil and allow to stand. Note the colour of any precipitate and note which carbohydrates do not react.

3 *Hydrolysis of a disaccharide* Dissolve a small portion of sucrose in $5 \, cm^3$ of 2M hydrochloric acid and heat in a water bath for 5 minutes. Neutralize the acid with 2M ammonia and repeat the test with Fehling's solution.

4 *Carbonyl derivatives* Carbohydrates will react with 2,4-dinitrophenyl-hydrazine but more useful derivatives are obtained by allowing the carbohydrate molecule to react with phenylhydrazine in a 1 : 3 ratio by moles.

Weigh accurately 0.2 g of glucose or fructose and 0.6 g of hydrated sodium ethanoate; mix with $4 \, cm^3$ of a 10 per cent solution of phenylhydrazinium chloride in water. Take care when handling the solution: it is harmful to skin and eyes.

Note the time and heat the mixture in a boiling water bath with occasional shaking. Record the time taken for crystals to appear and make a drawing of their crystalline appearance using a microscope. The appearance of the crystals and the time they take to appear are characteristic of the various carbohydrates.

5 *Polarized light* In Topic 7 you investigated the effect certain crystals had on polarized light. You are now going to find out if *molecules*, as distinct from a *crystal*, can show a similar effect. To obtain separate molecules in a non-crystalline arrangement, we are going to use a solution.

You will also have access to a polarimeter. Figure 11.9 shows how this instrument is constructed.

If solutions of carbohydrates are not available prepare them by dissolving 15 g in $100 \, cm^3$ of warm water. Half fill the specimen tube from the polarimeter. *Without* placing the specimen tube in position adjust the polarimeter by rotating the centre of the analyser until, on looking through the analyser and polarizer, you see that the source of light is extinguished. Note the position of the pointer on the scale. Put the specimen tube in position and look through the instrument once more. Do you have to alter the setting of the analyser to extinguish the light, and if so, by how much and in which direction, clockwise or anticlockwise? Now fill the specimen tube so as to double the length of liquid through which the light passes. Is a further adjustment of the analyser necessary for extinction?

As an additional experiment if you have time, you can investigate the hydrolysis of sucrose, using the polarimeter. Dissolve 100 g of sucrose in $40 \, cm^3$ of hot water and leave for 15 minutes to cool and clear. Add $5 \, cm^3$ of concentrated hydrochloric acid, mix well, and pour into the polarimeter tube. Take a reading of the setting of the analyser (α_t) and note the time (t). Take further readings every 5 minutes for about 45 minutes until the readings do not change in value (α_0). Plot a graph of ($\alpha_t - \alpha_0$) against (t) and comment on the shape of the graph.

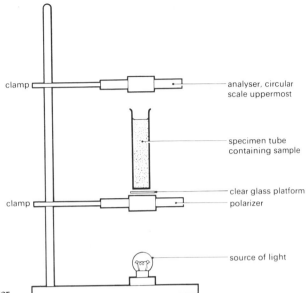

Figure 11.9
A simple polarimeter.

Reactions of the carbohydrates

1 *Dehydration* Concentrated sulphuric acid is a dehydrating agent and this reaction may remind you of the empirical formula of the carbohydrates.

$$C_6H_{12}O_6 \xrightarrow{\text{concentrated } H_2SO_4} 6C + 6H_2O$$

Concentrated sulphuric acid is also an oxidizing agent and the reaction is more complex than the simple equation given.

2 *Fehling's solution* Fehling's solution can be used to test for reducing sugars. The monosaccharides fructose and glucose react readily but the di-saccharide, sucrose, and the polysaccharide, starch, do not react because the carbonyl functional groups have been used to link their monosaccharide residues together. Thus sucrose and starch have to be hydrolysed to their constituent monosaccharides before reaction occurs with Fehling's solution.

$$\text{sucrose} + \text{water} \xrightarrow{\text{dilute HCl}} \text{glucose} + \text{fructose}$$

$$\text{starch} + \text{water} \xrightarrow{\text{concentrated HCl}} \text{glucose}$$

ethanoic anhydride. Each glucose monomer reacts on average with 2.4 to 2.9 ethanoyl groups. The fibres dissolve in the reaction mixture and are recovered by dilution.

The relative merits of various textile fibres are compared in Topic 17.

BACKGROUND READING 2
The sweetness of sugar

For all the flavours that we can distinguish in our foods we are indebted mainly to our sense of smell because it seems that our sense of taste is restricted to four sensations: salt, acid (or sour), sweet, and bitter.

The sensation of sweetness has for a long time been associated with a range of naturally occurring carbohydrates, and nowadays also with a small group of synthetic compounds. Honey as a source of sweetness must have been known in the Stone Age because a cave painting in Spain records a person robbing a wild bee's nest (figure 11.10).

Figure 11.10
Neolithic honey gathering.
From HERNANDEZ-PACHECO, FRANCISCO, Bulletin de Real Sociedad Espagnola, *Madrid, 1921, pp. 62–67.*

And sucrose from sugar cane has been known in the East for a very long time, being reported by an officer of an invading European army in the fourth century B.C. Sugar beet as a source of sucrose came into prominence during the Napoleonic Wars when in 1811 France was cut off from her traditional sources of cane sugar. In the last hundred years two synthetic agents have come into prominence, saccharin, discovered in 1879 and cyclamate, synthesized in 1937.

Different compounds have very different degrees of sweetness (see table 11.1) and this accounts for some of the success of the synthetic sweetening agents. Both saccharin and cyclamate were synthesized as part of research pro-grammes quite unrelated to sweetness and owe their discovery to unhygienic laboratory practice. The laboratories became contaminated, and this was noticed because of an unexpected intense sweetness. The source of sweetness was soon traced to the appropriate compound.

Compound	Source	Relative sweetness
Sucrose	cane and beet	1
D-fructose	fruit	1.7
'Invert sugar'	honey	1.2
D-glucose	fruit	0.5
D-lactose	milk	0.2
'Corn syrup'	corn starch	0.3
Saccharin	synthetic	300–550
Cyclamate	synthetic	30 +

Table 11.1
Relative sweetness of various compounds.

Cyclamate quickly became a popular additive to foodstuffs after 1950 because it lacks the bitter after-taste associated with saccharin and is not destroyed by heat. However, the wisdom of permitting cyclamate to be used as a food additive was questioned when it was found that a small proportion of people converted cyclamate to cyclohexylamine, and large doses of cyclohexyl-amine, when injected into rats, caused chromosome damage. Cyclamate was finally banned in the United States in 1969 after research reporting that rats given large daily doses of a mixture of cyclamate and saccharin developed bladder tumours. Within a week six governments around the World banned the use of cyclamates. Although the doses given to rats were very much higher than might be taken by humans, the use of cyclamates was banned in the United States because a part of that country's food law (the Delaney Amendment) says that if any substance is found to cause cancer in any animal, no matter what the dose level used, that substance shall not be used in food for humans.

You may well wonder why saccharin was not banned as well. The fact is that saccharin has been used for over 80 years by many people. including

Once a sufferer has been diagnosed as a diabetic, he or she has to maintain a near normal blood glucose level. Here too, simple colour tests are available which enable patients to check their own levels. Thus, a diabetic may begin to care for himself or herself whilst leading a full and active life.

11.5
SURVEY OF REACTIONS IN TOPIC 11

The chart below summarizes the main reactions that you have met in this Topic. You should copy the chart into your notebook and add appropriate details about the reagents, their chemical nature, and the types of reaction involved. It may help you to learn the material if you make several charts, one for each particular type of information.

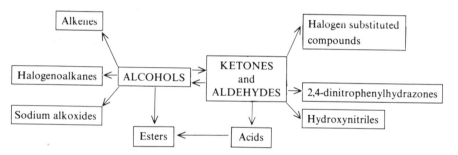

Draw up similar charts of your own for the reactions and properties of phenol and carbohydrates.

SUMMARY

At the end of this Topic you should:

1 understand, and be able to use, the systematic methods of nomenclature used for the carbon compounds that have been described in the Topic;

2 be familiar with some further organic practical procedures;

3 know the chemical reactions of the various classes of compounds that are given in the parts of the Topic headed 'Reactions of...' and summarized above;

4 be aware of the method of manufacture of phenol, and of its importance in the manufacture of synthetic polymers and fibres;

5 understand what is meant by the term 'chirality', and the type of molecular structure that gives rise to chiral compounds;

6 understand the meanings of the terms enantiomer, optical activity, dextrorotatory, and laevorotatory;

7 have some awareness of the matters discussed in the Background reading on diabetes and on sweetness.

PROBLEMS

1 Draw structural formulae for the following:

a 2-methylpropan-2-ol

b 2,2-dimethylpropan-1-ol

c pentan-3-one

d 3,4,4,5-tetramethylcyclohexa-2,5-dieneone ('Penguinone')

2 Name the following compounds:

a $CH_3CH_2CH_2CH(OH)CH_2CH_3$

b $CH_3CH\!=\!CHCHClCH_2OH$

c $CH_3COCH_2CH_3$

d $CH_3CH(OH)CH_2CH_2CHO$

3 The following substances all have the same molecular formula $C_7H_{15}OH$.

$$\text{A } CH_3-\overset{\displaystyle H}{\underset{\displaystyle CH_3}{\overset{\displaystyle |}{\underset{\displaystyle |}{C}}}}-\overset{\displaystyle CH_3}{\underset{\displaystyle H}{\overset{\displaystyle |}{\underset{\displaystyle |}{C}}}}-\overset{\displaystyle H}{\underset{\displaystyle H}{\overset{\displaystyle |}{\underset{\displaystyle |}{C}}}}-\overset{\displaystyle H}{\underset{\displaystyle H}{\overset{\displaystyle |}{\underset{\displaystyle |}{C}}}}-OH$$

$$\text{B } H-\overset{\displaystyle CH_3}{\underset{\displaystyle CH_3}{\overset{\displaystyle |}{\underset{\displaystyle |}{C}}}}-\overset{\displaystyle H}{\underset{\displaystyle H}{\overset{\displaystyle |}{\underset{\displaystyle |}{C}}}}-\overset{\displaystyle OH}{\underset{\displaystyle CH_3}{\overset{\displaystyle |}{\underset{\displaystyle |}{C}}}}-CH_3$$

$$\text{C } CH_3-\overset{\displaystyle H}{\underset{\displaystyle CH_3}{\overset{\displaystyle |}{\underset{\displaystyle |}{C}}}}-\overset{\displaystyle OH}{\underset{\displaystyle CH_3}{\overset{\displaystyle |}{\underset{\displaystyle |}{C}}}}-C_2H_5$$

$$\text{D } CH_3-\overset{\displaystyle OH}{\underset{\displaystyle CH_2CH_3}{\overset{\displaystyle |}{\underset{\displaystyle |}{C}}}}-CH(CH_3)_2$$

a Which substance is identical with **C**?

b Which substance, if any, is a secondary alcohol?

c Which substance is 2,4-dimethylpentan-2-ol?

d Which substance could be oxidized to an aldehyde?

e Which substance could be oxidized to a carboxylic acid containing the same number of carbon atoms?

4 In an experiment 3.7 g of butan-1-ol were heated with excess potassium bromide and concentrated sulphuric acid. The main product of the reaction was obtained in a yield of 2.74 g after purification.

a Name the product and write an equation for the reaction.

b Calculate the percentage yield of the product by comparing the actual yield with the maximum theoretically obtainable.

ei When an alcohol containing a large proportion of ^{18}O atoms is reacted with a carboxylic acid, the ^{18}O atoms are found in the organic product and not in the water produced during the reaction. State briefly what can be deduced about the mechanism of the reaction.

ii Draw a diagram of the structural formula of the probable product of the reaction between thioethanol and ethanoic acid.

iii What would be a suitable catalyst for this reaction?

9 Phenol vapour reacts with hydrogen, when the mixture is passed over a heated nickel catalyst, to give a compound **v** with the molecular formula $C_6H_{12}O$. **v** is readily oxidized by $K_2Cr_2O_7$—H_2SO_4 to **w**, $C_6H_{10}O$, which can be oxidized further, under more powerful conditions, to **x**, $C_6H_{10}O_4$. The compound **v**, when heated with moderately concentrated sulphuric acid, gives **y**, C_6H_{10}, which reacts with bromine to give **z**, $C_6H_{10}Br_2$.

Write structures for **v**, **w**, **x**, **y**, and **z**, and comment on these changes.

10 A substance, **A**, had the molecular formula C_3H_7Br. After boiling with aqueous sodium hydroxide a compound, **B**, of molecular formula C_3H_8O, was formed. On oxidation **B** formed **c**, of molecular formula C_3H_6O, which gave a precipitate with 2,4-dinitrophenylhydrazine, but had no reaction with Fehling's solution. Name and give the structural formulae of **A**, **B**, and **c**.

11 Triiodomethane, CHI_3, is formed as a yellow precipitate when a mixture of iodine and potassium hydroxide reacts with any of the compounds C_2H_5OH, $(CH_3)_2CHOH$, $(CH_3)_2CO$, $CH_3COC_2H_5$, and $C_6H_5COCH_3$. Triiodomethane is NOT produced when iodine and potassium hydroxide are added to any of the compounds CH_3OH, $HCHO$, $CH_3CH_2CH_2OH$, $(C_2H_5)_2CO$, and C_6H_5CHO.

a According to these results, what structural feature or features must be present in a compound if the action of iodine and potassium hydroxide on it is to produce triiodomethane?

b Suggest the stages which may be involved in the reaction.

12 A substance, **A**, had the molecular formula $C_4H_{10}O$. On oxidation it gave **B** (C_4H_8O) which gave a precipitate of copper(I) oxide with Fehling's solution.
On passing the vapour of **A** over heated silica it formed **c** (C_4H_8).

c reacted with hydrogen iodide to form D (C_4H_9I). D after hydrolysis and then oxidation formed E which gave a precipitate with 2,4-dinitrophenylhydrazine, but had no reactions with Fehling's solution. Name and give the structural formulae of A, B, C, D, and E.

13 Describe and attempt to classify (*i.e.* clearly indicate the reaction type and the kind of reagent) the reactions of the —OH group in a variety of organic compounds. Refer to specific reactions in your answer, giving structural formulae where possible.

14 Write a short essay to compare and contrast the reactions of the $\diagdown C = O \diagup$ group with $\diagdown C = C \diagup$. Use equations to help draw attention to the main points of similarity and difference and classify the reactions according to their mechanistic pattern.